Social Bodies

Social Bodies

Edited by
Helen Lambert and Maryon McDonald

Berghahn Books
NEW YORK • OXFORD

Published in 2009 by

Berghahn Books

www.berghahnbooks.com

©2009 Helen Lambert and Maryon McDonald

Library of Congress Cataloging-in-Publication Data

Social bodies / edited by Helen Lambert and Maryon McDonald.

    p. cm.

Includes bibliographical references and index.

ISBN 978-1-84545-553-8 (alk. paper)

1. Body, Human--Social aspects--Case studies. I. Lambert, Helen, 1960- II.
McDonald, Maryon.

GN298.S63 2009

306.4--dc22

                              2008052537

British Library Cataloguing in Publication Data

A catalogue record for this book is available
from the British Library

Printed in the United States on acid-free paper.

ISBN 978-1-84545-553-8 (hardback)

CONTENTS

INTRODUCTION

Helen Lambert and Maryon McDonald

That apparently singular object, the human body, has been the focus of much discussion in both academic and public quarters recently, with striking claims being made for fundamental changes in our attitudes towards the body and its parts as a result of technological developments. This volume seeks to examine some of these changes whilst at the same time interrogating the object that seems to be at the heart of all the fuss. Certainly there is a sense, at least in European and American contexts, that in concert with developments in genetics and biomedicine, both treatment of the body and attitudes to it have changed and that formerly taken-for-granted aspects of our corporeal selves can no longer be relied upon. This much has been obvious in the proliferation of 'ethical' concerns, in press headlines and in social science texts. Even before this recent surge of public attention, however, there was a growing interest in 'the body' among academicians, and numerous social science texts have tried to pinpoint reasons for this (see Van Wolputte 2004). Late capitalism, feminism and the biotechnology industry, with its apparent 'remaking of life and death' (Franklin and Lock 2003), have been favourites in this literature.[1]

In social anthropology, the main discipline that informs this volume, a growing interest in the body initially meant a shift from the body as a taken-for-granted aspect of life to the body as explicit object, apparently singular and available for study.[2] However, as anthropological studies of 'the body' have proliferated, this once discrete object has analytically dissolved and the relationalities and realities involved have appeared to multiply. To look more carefully at these themes, the Royal Anthropological Institute, in conjunction with the Leverhulme Trust, launched a series of workshops. The first of these, on Genomics, was held in April 2005 and began an examination of the relationalities of the new bodies presented by genomic science.[3] The second, with the very general title of 'Social Bodies', was held in September 2005 and most of the chapters of this volume were first presented there.

By 'social bodies', we did not have in mind older notions of the social body that we might find as one of the two bodies in the work of Mary Douglas (1970), for example. The phrase is not meant to suggest a representational reality constraining the way the physical body is perceived. Nor is it one of the three or even five bodies since delineated by Scheper-Hughes and Lock (1987), O'Neill (1985) and other anthropologists and sociologists (see Csordas 1999 and Van Wolputte 2004 for some of this literature). All this pioneering work has been essential to our own but this volume does not aim to look either at 'the body' as object or at the 'social(ised) body' as delineated by any of the literature cited; rather, it takes up – and pursues in a different direction – the non-dualistic approaches characterized by Lock as 'making the body social' (Lock 1993: 135) whilst at the same time critically examining the notion of 'embodiment' where these approaches have generally ended up.

The juxtaposition of 'bodies' and the 'social' in our title was intended, in part, to prompt reflection on what happens to sociality in the face of new types of transaction (organ transfer, forensic identification, stem cell therapy, new reproductive technologies) that focus squarely on elements of the corporeal. Much of the anthropological literature on the body now relies on a phenomenological approach in order to emphasize embodiment, thereby producing a focus on the body-as-self (Csordas 1994; Sharp 2000: 289). When deployed to address new medical technologies and transcultural movements in commodified human tissues and products, this approach inevitably construes body fragmentation and commodification occurring in scientific and medical practices as signalling worrying disturbances in the ability of human subjects to integrate their bodies, selves and persons. In this volume, by contrast, we do not take for granted an incorporated personhood that is coterminous with the boundaries of the human body. Rather we seek, partly in contradistinction to the recent emphases in social science analyses on the commodification and objectification of bodies and their parts (see Sharp 2000; Scheper-Hughes and Wacquant 2003), to consider the extent to which bodies and their elements are themselves 'social' and whether we can identify aspects of sociality that are inherent to human bodies.

One such aspect concerns the relations between bodies and what are perceived to be their biological components. Much of the anthropological literature emerging from studies based in Europe and the United States overtly or implicitly assumes that the scientific knowledge and technical operations that enable, for example, cells to be cloned or organs to be transplanted, necessarily give rise to new social formations and cultural concerns. As many authors have noted, however (for example Lock 2002; Peers 2003; Fairhead, Leach and Small 2006), and as several of our chapters demonstrate, human matter when divorced from its source is frequently a focus of concern and elaboration. In many parts of the world, common media for sorcery practices have been human hair and nail clippings, while human placentas and blood are likewise frequently the focus of ritual procedures designed to associate their source bodies with particular localities or to protect against the possible harms that may accrue to these bodies if their products are illicitly manipulated or accidentally disturbed (Aijmer 1992; Lambert 2000a). These practices point to an enduring, if generally implicit, understanding that bodily components retain a substantive connection with their source after separation. This understanding is necessarily coloured by the broader sets of relations in which it exists. We see this in responses to clinical and public health activities in colonial and postcolonial settings such as India and Africa, for example, where interventions that require breaking of the skin (as in immunization) or the collection of blood have often given rise to conflict with state authorities who are generally suspected of having nefarious or exploitative purposes (e.g. Vaughan 1991; Arnold 1993; Fairhead; Leach and Small 2006; Geissler and Pool 2006). Comparative analysis can help us to discern both similarities and discontinuities between such attitudes and, for example, the perceptions of contemporary US organ donors towards their new body parts (Sharp 1995) or the interpretation of 'biobanking' activities among potential participants. In a post-Alder Hey world in the UK, the regulatory path to the Human Tissue Act (2004) has been strewn with the elisions of parts, bodies and individual persons. At the same time, a distinction between body and non-body (such that a tissue slide, for example, if not a medical record, was body part and therefore person) was one important but difficult issue for which resolution was sought in this Act (Parry 2005; Gere and Parry 2006). Some of the chapters of this volume evoke not dissimilar tensions between 'data' and persons.

Science and the Social

Given that the origins of new transactions of human matter across body boundaries lie in scientific developments and technological innovation, especially in the medical realm, the juxtaposition of 'bodies' and the 'social' inevitably evokes many of the issues that were previously characterised by social scientists as concerning 'science and society'. Each of these categories has a history; they have ethnographic reality and are still frequently, in Latour's terms, 'co-produced' (Latour 1993). In the 1960s, the 'science and society' dyad was invoked to codify sociopolitical concerns emerging primarily from the harnessing of nuclear energy for both military and non-military purposes; in the post-Cold War, postmodern period it declined in significance as an analytic focus, but has re-emerged as worthy of examination in more recent years. A division between the social and the scientific is one that can have ethnographic salience; in other words, it is sometimes part of the theory through which people constitute themselves and their world, rather than being derived from our own analytical toolkit as anthropologists (Lock 2001; McDonald 2006, and forthcoming). This is not to suggest that 'science' and 'society' each retain their integrity ethnographically in the terms of the old 'science and society' debates: evidence from empirical studies demonstrates that scientific knowledge in effect dissolves when examined in use within specific public contexts (Lambert and Rose 1996). At the same time, as this volume suggests, the duality has ethnographic salience in that bodies are seen to shift definitionally between scientific and social domains, between scientific practice and the domain of social persons, whether separated or brought together by the protagonists themselves.

It is of particular relevance to the chapters in this volume and to our focus on bodies that the source and focus of social concerns about 'science' have themselves shifted. From roughly the 1960s to the last decade of the twentieth century, concerns about the relationships between 'science', always implicitly standing for the natural sciences, and society evolved primarily in response to developments in physics. Relations between the natural sciences and the social came to be delineated in two largely separate bodies of research and writing. The field generally characterized as concerned with the 'public understanding of science' drew primarily on a deficit model of popular ignorance about scientific knowledge (see Irwin and Wynne 1996 for a summary and critiques of this position), while sociological analysis of science as itself a socially constructed activity flourished under the rubrics of 'STS' (science, technology and society) studies and 'SSK' (sociology of scientific knowledge). These dual approaches in the social sciences to the relationship between scientific activities and knowledge, on the one hand, and the social milieux within which these activities and knowledges are seen as situated, on the other, have largely been sustained, while public interest in these issues has fluctuated. Towards the end of the twentieth century and into the new millennium, however, the focus of both academic and public concern has, with the development in particular of molecular genetics, shifted from the physical to the biological sciences and their technological applications. In seeking to understand the sociality of human bodies, many of our contributors perforce examine the role of particular biological sciences and their practices and effects on particular kinds of bodies and body parts.

We live in a world now where the biosciences offer important languages in which bodies are understood and communicated; however, *Social Bodies* is not about the social authoring of biology in the sense of bioengineering nor about the social authoring of the discipline of biology. Rather this volume takes for granted the analytical point that scientific practices, while they may imagine themselves to be ideally shorn of the social and the personal, have an inherent sociality. This point has been well established empirically; in an analytical sense, scientific practice, inside or outside the laboratory, is necessarily social and political.[4] At the same time the biomedical sciences, for example, attempt in their anatomical training and practice to construct and examine momentarily asocial bodies in which the legalistic practice of anonymization – derived from the 'social' as it reappears in the form of '(bio)ethics' – can then appear to collude.[5] The desire of others to rename, to make social, the bodies that are thus implicated in scientific practice can sometimes appear, as suggested by the contributions of Peers and Strathern to this volume, to be merely a cultural perspective on an underlying object – the material body or body part – which science has already properly grasped. Other chapters suggest how the socially situated nature of scientific practices can nevertheless facilitate as much as deny social relations to their object.

Bodies are Social: The Chapters of This Volume

The conjunction of *Social Bodies* emphasizes therefore that, in an analytical sense, all bodies are social, including the bodies that the natural sciences have constructed in the self-perceived context of a knowledge ideally free of the social and political. The contributions to this volume demonstrate in a variety of ways how – contrary to the assumptions of the various sciences (forensic medicine, transplant surgery, immunology, epidemiology, public health) whose practices form the starting point for some of our case studies – acts of bodily fragmentation, dismemberment, transfer, reassignment or transformation are never confined solely to the biologically functional in their effects but inevitably entail the reformulation, reconstruction or reestablishment of social relations between persons and between human groups. This is not to say that scientific practices are always directed towards severing what are, in their terms, the material from the social elements of human bodies (Peers). A notable feature of this collection is its demonstration of the ways in which particular scientific practices and knowledges can also serve, on the contrary, to integrate bodies with social identities, for example by reuniting corpses with names and identities (Petrović-Šteger, Robb) or museum specimens consisting of human remains with source communities (Peers). This movement of human corporeality between what are perceived to be distinct social and material dimensions, and the shifting position of a variety of scientific knowledges and activities in constraining or sustaining such movement, is a pervasive theme of the volume.

In the first chapter, by Sharon R. Kaufman, Ann J. Russ and Janet K. Shim, the relationality of bodies is shown to take on an ethnographic reality in which living, but sick, bodies become an 'ethical field' through which new socialities are constructed or existing ones reinforced. Kaufman, Russ and Shim's US study reveals an inequality of obligation between the living donors and recipients of kidneys, whereby a determination to give as an expression of caring must overcome a reluctance to receive. The sense of being 'like family' that motivates the desire to give (often, but not always, to close kin) among these living donors is particularly striking when contrasted with the perfect reciprocal seen in anthropological work on cadaveric organ donation. This was highlighted for us at the original 2005 workshop by comments from the American anthropologist Lesley Sharp. Her work (1995, 2005, 2006) in particular has documented the ways in which relatives of deceased donors in the US and the unrelated recipients of these donors' organs strongly desire a connection that they construe as a form of kinship. Thus in Kaufman et al.'s study a sense of relatedness motivates the transfer of a body part (often, though not exclusively, to an older relative) whereas in cadaveric donation, the transfer of a body part (from a theoretically anonymous donor) prompts a sense of relatedness. Far from indicating some fundamental inconsistency, these contrastive cases demonstrate that for ordinary Americans, social connection is a necessary component of corporeal transactions between discrete bodies. In living donation, donors and recipients inhabit, in Kaufman et al.'s terms, different ethical spaces mediated by biomedicine, which provides the currency through which to express love and care. Where relatedness can be maintained via organ transfer, as for Kaufman et al.'s informants, the inherent sociality of this bodily transaction seems to remain implicit, unremarked upon, even while it is precisely this sense of relatedness that creates a moral obligation to donate. Conversely, where relatedness is disconnected through death and the transfer of deliberately anonymized body parts between strangers, as in cadaveric donation, recreating relatedness seems to become (at least for these American donors and recipients) an imperative.

Kaufman, Russ and Shim and others' work on organ donation serves to demonstrate the inherent sociality of body components in another way. Work on cadaveric donation in the US has highlighted a contrast between the dissolution and disappearance of the cadaveric donor body that is assumed by transplant professionals to occur following its anonymization and fragmentation, and the deliberate re-personification of donated organs by both bereaved relatives and living recipients (see especially Sharp 1995, 2006). Health professionals' determination to anonymize cadaveric donors (on ethical grounds) conflicts directly with donor kin's determination to name them (understood as an essentially moral act). The divergence between official anonymisation procedures and the relatedness potentially created through donation is seen too, in a different configuration, in studies of UK ova donation (e.g. Konrad 2005). This conflict can be seen as a definitional struggle of some import: whether to assign the act of organ or tissue donation to the category of the biological – in the sense of that discipline of scientific enquiry that treats the corporeal as purely material – or the social. In her chapter, Laura Peers argues that human remains found in museum collections similarly swing between these two poles, being regarded by different parties as scientific specimens or as ancestors. These shifts of meaning between bodies and bodily materials as physical matter and as social beings are to be found in most of the chapters making up this volume.

These definitional conflicts are played out in different ways in the following chapters. In the chapter by Petrović-Šteger, anatomical science is called on to construct bodies from human remains so that social identities can be reconstituted. Petrović-Šteger shows how (like Sharp's relatives of organ donors) anatomy as employed in forensic science moves beyond a notionally asocial materiality to seek thereby the determination of a corpse's identity and the reconstitution of corporeal fragments into named persons. In Petrović-Šteger's contribution, the salvaging of material bodies through scientific knowledge contrasts with the use she outlines of biomedical information by kinspersons and other participants to reclaim the sociality of those bodies. Clearly the natural and medical sciences are not always positioned in support of a 'pure material substance' reading of the human body in these cases. The forensic scientists and associated professionals in Petrović-Šteger's Serbia are, by re-personalizing and sourcing body parts, doing the opposite of most European and American organ transplant teams.

The status and meaning of human remains are the focus also of the subsequent two chapters, by Peers and Robb. As in Petrović-Šteger's example, the cases they describe provide the means to illuminate the broader problematic of the character of bodily sociality. Peers's critique of UK scientists, who regard museum specimens composed of indigenous human remains as purely biological material that must be retained for the sake of research, pinpoints a theme of central importance to science-society debates, that of the relationship between trust and expertise. Her analysis illustrates how social relations between communities of origin continue to inhere in human remains that have historically been dissociated by their collection and removal from these communities. It further considers how contemporary social relations between these groups of origin and those who collected the remains are both affected by and affect the status of these remains. The contributions by Petrović-Šteger and Peers shift our focus from consideration of the relations between individual bodies as expressed or created through the transfer between them of body parts, to the relations among groups (whether seen in terms of ethnicity, locality, race or otherness). Peers's discussion of an ongoing conflict between English museum curators and scientists who use their collections as data, and the indigenous peoples who lay claim to the objects in these collections composed of (relatively recent) human remains on grounds essentially of relatedness, foreshadows Strathern's reflection on these disputes in her contribution. The methodological and political difficulties inherent in attempting to define and particularize relations between individual human remains and social groups are illuminated in another direction by Kaestle's (2003) discussion of the pitfalls in assigning membership of ancient human remains through DNA analysis to particular contemporary human groups, who thereby become putative descendants and representatives of origins with which they may or may not identify.

Robb's chapter employs as a case study one such particularly well-known archaeological find – the 'Ice Man' – and shows how, when we move beyond the skeletal remains to soft tissue, we also move into the invention of names, relations, lifestyle and even nationality. His chapter illustrates well some of the very real problems of the archaeological approach of combining DNA analysis with selected ethnological assumptions in order to interpret ancient remains. Robb's account of this famous discovery and its subsequent treatment by both archaeologists and the media demonstrates the need for 'an updated version of ethnographic analogy' (Silverman 2003: xii) to enable a more sophisticated and flexible use of the possible social arrangements documented ethnographically in order to illuminate archaeological findings. Robb's 'Ice Man' also illustrates how tangibly soft human remains in Europe readily become bodies and persons in a unity of theory and observation. It becomes very clear here that, conceptually and socially, there is no non-relational body.

In the last two chapters, by Aparecida Vilaça and Marilyn Strathern, we are asked to confront common assumptions that may get in the way of any analysis that claims to focus on bodies. In recent years, the idea of 'embodiment' has been used by many social scientists to suggest a material antidote to an apparently idealist social science, to go beyond an analytical dichotomy of mind and body, or to inject a sense of 'agency' and 'process' into a world of structure and discourse (see Jackson 1989; Toren 1998; and Csordas 1999 for an overview). Whatever one thinks of such dualities, the 'embodiment' paradigm (notably in medical anthropology and the work of Csordas 1994, for example) has achieved acknowledged analytical success in areas where other approaches have been deemed wanting, and it is an approach that is rigorously both corporeal and social in its emphases. At the same time, however, it carries echoes of the assumptions and dualities it seeks to go beyond as it patches together natural science and phenomenology. It is an approach that, as we have suggested above, tends to rely on assumptions about 'the body' as an individual, stable substrate coterminous with the boundary denoted by the human skin and housing a 'subject' at the centre of relations. Such an approach, the last two chapters of this volume suggest, can be understood as part of a Euro-American emphasis on diverse perspectives underpinned by a unified material reality that is at odds with the 'perspectivism' of Amazonia or aspects of the complex relationalities of Melanesia. In both of these last contexts, singular or stable 'bodies' are inherently and inseparably social rather than being their anterior and individuated base. Vilaça argues that the phenomenological approach is based implicitly on a Euro-American concept of the person that presumes a universality of human nature but a variety of cultures, whereas Amerindian notions of personhood posit the same culture and multiple natures (cf. Viveiros de Castro 1998).

Vilaça's demonstration that Amerindians take a capacity for metamorphosis between humans and other animal species to be a fundamental feature of humanity, so that procedures for well-being are focused on constraining this mutability, resonates with findings in other cultural contexts. In South Asia, for example, human caste groups can be seen as constituting discrete 'species' or 'kinds' (Hindi *jati*) (Marriott and Inden 1977). Marriott famously argued that in this setting, Schneiderian 'substance' and 'code for conduct' are inseparable so that action and ingestion are transformative, and he characterized the substantively interconnected nature of persons in India by describing them as 'dividuals'. Bodily health and social standing are thus implicated in appropriate social practice, and numerous other authors have described the 'ecological' character of these understandings of bodies and persons as constituted by continual flows of substance (Marriott 1976; Daniel 1984; Zimmerman 1988; Busby 1997; Lambert 2000a). Among Hindus, restrictions on these flows of substance (food, water, milk, blood) between persons can in turn be seen as ways of differentiating kin (those who are substantively alike) from non-kin (those who are substantively dissimilar) (Lambert 2000b).

Returning to South America, Vilaça describes the requirement for Amerindian kinspersons of the dead to accept that the deceased are ex-human so that they can be relinquished. This contrasts markedly with American cadaveric organ donation wherein the part is held to preserve the human, social identity – the personhood – of the donor (Sharp 1995). Strathern in turn presses the point of cultural particularity and projection further by arguing that as the notion of the body as a complete, totalized entity is a Euro-American projection, the preoccupation over fragments results from the fantasy of such a whole. Her contribution observes that bodies are interdependent, that they comprise a field of relations, and that the 'fractal persons' of Melanesia and Rwanda (for example) are parts of relations, not parts of (potentially complete) bodies, insofar as bodies always exist in relation to one another with flows between them.

Metaphors and practices of 'donation' in organ, blood, tissue or gamete transplantation suggest connections between persons that can make the relationalities of Melanesia, for example, seem less exotic. The bodies of Kaufman et al.'s chapter on the USA certainly comprise the specific historical actions of others, which those of Melanesia appear to do in a more routine way (Strathern 1988). In the US, a belief in moral commitment between persons can sit side by side with a biomedical belief that bodies are, in their most elementary form, material entities that can be turned into abstracted body parts. As Kaufman et al.'s chapter makes clear, being confronted with end-stage kidney disease in the US increasingly means existentially to 'engage the sociality of the body' (p. 21); moral relations are made corporeal, and the figure of the body as relational social substance is then foregrounded.

Social Substance

Taken together, these chapters essentially concern a set of problematics around two kinds of relatedness: between (whole) bodies and groups of bodies, and between parts of bodies and their source bodies. The contributions in this volume suggest that in this sense, bodies exist only where there are relationships; personhood and relatedness (or social substance) emerge as being essential to the 'bodiliness' of human material. Bodies do not express, incorporate or embody the social. Rather, they *are* inherently social. This is not naïve social constructionism nor an analytical re-erection of a division between the material and the social. Rather, bodies are social in their materiality.

In this interpretation, we depart analytically from Lambek's (1998: 104) claim that bodies, here contrasted with minds, are of relevance anthropologically only when they are related to other significant categories (person, self, mind etc.), for 'anything less is simply biology'. This analysis in effect leaves intact the notion of an isolable, asocial materiality which the biosciences have helped to construct – but which soon finds the 'social' flooding back (McDonald, forthcoming) – as bioethics, as scandal and controversy, as relatedness, as social persons. The approach of Lambek just mentioned is one that we have wished to make not an analytical standpoint but a matter of ethnographic interest – part of the theories through which people constitute their worlds. In some instances, the materiality in which the inherently social body has notionally disappeared may be the artefact of scientific practice, which that same practice can also help to repair, but in another (Vilaça) it is the achievement of funerary ritual. *Bodies*, as several of our chapters tellingly suggest, may be distinguished from human remains or corpses. Ethnographically, 'bodies' can seem to disappear into 'human remains' – the latter being, it seems from the evidence presented in this volume, not a euphemism but an apposite reference to what is left when the body has gone. Forensic science is called on to construct bodies from human remains and this is done not solely by reconstituting their parts but by identifying the clothing that once adhered to them and reconnecting them via DNA analysis to their names and their kin. We see, too, acts to reclaim the person from the fragmented body and to reconstitute the body in order to reclaim the person. These operations testify to the relationship of identity between bodiliness as at once material and social. The very language of this volume further exemplifies the simplicity of this point. Someone who lacks social standing in a given societal milieu is, in English, described as a 'nobody' – someone literally without a body. A key aspect of the sociality of bodies resides in the point that a 'body' is everywhere constituted by social connections, with the connections – and thus the bodies – varying between England (Strathern 1981; Edwards 2000), Melanesia (Strathern 1988), West Africa (Piot 1999), Corsica (Candea 2004) and so on. Such examples and others in this volume suggest a fruitful line of comparison for examining the starting point of this introduction: recent controversies concerning 'the body'. If we imagine that these social connections are simply connecting individual bodies, or that underneath all the discourse of the social lies 'the body', then this volume should cause us to reconsider. The chapters of this volume move us ethnographically from 'the body' to a plurality of 'bodies' and they move us analytically from 'the body' to 'embodiment' and beyond, to 'social bodies'. At the same time, they provide some new ethnographic insights into bodies as sociality, as well as the material of a potentially powerful analytical reconsideration of notions of embodiment.

Notes

1. For surveys of some of this material, see also Martin (1992), Csordas (1999), Fraser and Greco (2005). Feminism figures prominently as an important source of critical interest in the body, with some authors looking back to Simone de Beauvoir, as further indicated by Bordo's analysis of the gendering of mind-body dualisms (Bordo 1993), for example, and Morgen's overview of the politics of the women's health movement in the USA (Morgen 2002). Moreover, kinship and gender studies by feminist scholars have laid much of the ground for later social studies of science and technology in their attention to the operations of the 'natural', the politics of reproduction and the gendering of scientific knowledge production. See Franklin's (1995) review and later work (e.g. Franklin 1997); also Haraway (1989, 1997) and Jordanova (1989), to name just a few examples in a huge and still growing literature.
2. Lock, in an early review of the field in anthropology, traces this shift in perspective to the late 1970s (1993: 134).
3. See Konrad and Simpson (2005) for an account of this workshop.
4. One of the pioneering ethnographic studies of scientific practice was Latour and Woolgar (1979); see also Latour (1987, 1989, 1993); Rabinow (1996); and Franklin's 1995 overview.
5. On some key aspects of medical training here, see Good (1994) and Sinclair (1997); for rather different aspects of anonymity, see Konrad (2005).

References

Aijmer, G. 1992. 'Introduction: Coming into Existence', in G. Aijmer (ed.), *Coming into Existence: Birth and Metaphors of Birth*. Gothenberg: IASSA.

Arnold, D. 1993. *Colonizing the Body: State Medicine and Epidemic Disease in Nineteenth-century India*. Delhi: Oxford University Press.

Bordo, S. 1993. *Unbearable Weight: Feminism, Western Culture, and the Body*. Berkeley: University of California Press.

Busby, C. 1997. 'Permeable and Partible Persons: A Comparative Analysis of Gender and Body in South India and Melanesia', *Journal of the Royal Anthropological Institute* n.s. 3: 21–42.

Candea, M. 2004. 'In the Know: Being and Not Being Corsican in Corsica', Ph.D. thesis. Department of Social Anthropology, Cambridge, UK.

Csordas, T. (ed.). 1994. *Embodiment and Experience: The Existential Ground of Culture and Self*. Cambridge: Cambridge University Press.

——— 1999. 'The Body's Career in Anthropology', in H. Moore (ed.), *Anthropological Theory Today*. Cambridge: Polity.

Daniel, E.V. 1984. *Fluid Signs: Being a Person the Tamil Way*. Berkeley and Los Angeles: University of California Press.

Douglas, M. 1970. *Natural Symbols*. London and New York: Routledge.

Edwards, J. 2000. *Born and Bred: Idioms of Kinship and New Reproductive Technologies in England*. Oxford: Oxford University Press.

Ewing, W.A. 1994. *The Body: Photoworks of the Human Form.* London: Thames & Hudson.

Fairhead, J., M. Leach, and M. Small. 2006. 'Where Techno-science Meets Poverty: Medical Research and the Economy of Blood in The Gambia, West Africa', *Social Science and Medicine* 63: 1109–20.

Franklin, S. 1995. 'Science as Culture, Cultures of Science', *Annual Review of Anthropology* 24: 163–84.

——— 1997. *Embodied Progress: a Cultural Account of Assisted Conception.* London and New York: Routledge.

Franklin, S. and M. Lock. 2003. 'Animation and Cessation: The Remaking of Life and Death', in S. Franklin and M. Lock (eds), *Remaking Life and Death: Toward an Anthropology of the Biosciences.* Oxford: James Currey, pp. 3–22.

——— (eds). 2003. *Remaking Life and Death. Toward an Anthropology of the Biosciences.* Oxford: James Currey.

Fraser, M. and M. Greco. 2005. 'Introduction', in M. Fraser and M. Greco (eds), *The Body: A Reader.* London and New York: Routledge.

Geissler, W. and R. Pool. 2006. 'Editorial: Popular concerns about medical research projects in sub-Saharan Africa – A critical voice in debates about medical research ethics', *Tropical Medicine and International Health* 11(7): 975–82.

Gere, C. and B. Parry. 2006. '"The Flesh Made Word": Banking the Body in the Age of Information', *Biosocieties* 1: 41–54.

Good, B. 1994. *Medicine, Rationality and Experience: An Anthropological Perspective.* Cambridge: Cambridge University Press.

Haraway, D. 1989. *Primate Visions: Gender, Race and Nature in the World of Modern Science.* New York and London: Routledge.

——— 1997. *Modest Witness@Second Millennium.* London and New York: Routledge.

Irwin, A. and B. Wynne. 1996. 'Introduction', in A. Irwin and B.Wynne (eds), *Misunderstanding Science? The Public Reconstruction of Science and Technology.* Cambridge: Cambridge University Press, pp. 1–17.

Jackson, M. 1989. *Paths toward a Clearing: Radical Empiricism and Ethnography.* Bloomington: Indiana University Press.

Johnson, M. 1987. *The Body in the Mind: The Bodily Basis of Meaning, Imagination and Reason.* Chicago: University of Chicago Press.

Jordanova, L. 1989. *Sexual Visions: Images of Gender in Science and Medicine between the Eighteenth and Twentieth Centuries.* New York: Harvester Wheatsheaf.

Kaestle, F. 2003. 'The Good, the Bad, and the Ugly: Promise and Problems of Ancient DNA for Anthropology', in A. Goodman, D. Heath and M.S. Lindee (eds), *Genetic Nature/Culture: Anthropology and Science beyond the Two-Culture Divide.* Berkeley : University of California Press, pp. 278–96.

Konrad, M. 2005. *Nameless Relations: Anonymity, Melanesia and Reproductive Gift Exchange.* New York and Oxford: Berghahn.

Konrad, M. and R. Simpson. 2005. 'Anthropology and Genomics: Exploring Third Spaces', *Anthropology Today* 21(4): 23–24.

Lambek, M. 1998. 'Body and Mind in Mind, Body and Mind in Body: Some Anthropological Interventions in a Long Conversation', in M. Lambek and A. Strathern (eds), *Bodies and Persons: Comparative Perspectives from Africa and Melanesia.* Cambridge: Cambridge University Press.

Lambert, H. 2000a. 'Village Bodies? Reflections on Locality, Constitution and Affect in Rajasthani Kinship', in M. Bock and A. Rao, *Culture, Creation and Procreation: Concepts of Kinship in South Asian Practice*. New York and Oxford: Berghahn Books, pp. 81–100.

——— 2000b. 'Sentiment and Substance in North Indian Forms of Relatedness', in J. Carsten (ed.), *Cultures of Relatedness: New Approaches to the Study of Kinship*. Cambridge: Cambridge University Press, pp. 73–89.

Lambert, H. and H. Rose. 1996. 'Disembodied Knowledge: Making Sense of Medical Science', in A. Irwin and B. Wynne (eds), *Misunderstanding Science? The Public Reconstruction of Science and Technology*. Cambridge: Cambridge University Press, pp. 65–83.

Latour, B. 1987. *Science in Action: How to Follow Scientists and Engineers through Society*. Milton Keynes: Open University Press.

——— 1989. *The Pasteurization of France*. Cambridge, MA.: Harvard University Press.

——— 1993. *We Have Never Been Modern*. New York: Harvester Wheatsheaf.

Latour, B. and S. Woolgar. 1979. *Laboratory Life*. London: Sage.

Lock, M. 1993. 'Cultivating the Body: Anthropology and Epistemologies of Bodily Practice and Knowledge', *Annual Review of Anthropology* 22: 133–55.

——— 2001. 'The Tempering of Medical Anthropology: Troubling Natural Categories', *Medical Anthropology Quarterly* 15(4) December: 478–92.

——— 2002. 'The Alienation of Body Tissue and the Biopolitics of Immortalized Cell Lines', in N. Scheper-Hughes and L. Wacquant (eds), *Commodifying Bodies*. London: Sage, pp. 63–91.

Marriott, M. 1976. 'Hindu Transactions: Diversity without Dualism', in B. Kapferer (ed.), *Transaction and Meaning: Directions in the Anthropology of Exchange and Symbolic Behaviour* (ASA Essays in Social Anthropology Vol.1). Philadelphia: Institute for the Study of Human Issues, pp. 109–42.

Marriott M. and R. Inden. 1977. 'Toward an Ethnosociology of South Asian Caste Systems', in K. David (ed.), *The New Wind: Changing Identities in South Asia*. The Hague: Mouton, pp. 227–38.

Martin, E. 1992. 'The End of the Body?', *American Ethnologist* 19: 121–40.

McDonald, M. 2006. 'Neo-nationalism in the EU: Occupying the Available Space', in M. Banks and A. Guingrich (eds), *Neonationalism in Europe and Beyond*. Oxford: Berghahn.

——— [forthcoming]. 'Introduction' to M. McDonald (ed.), *Languages of Accountability*. Oxford: Berghahn.

Morgen, S. 2002. *Into Our Own Hands: The Women's Health Movement in the United States, 1969–1990*. New Brunswick: Rutgers University Press.

O'Neill, J. 1985. *Five Bodies: The Human Shape of Modern Society*. Ithaca, NY: Cornell University Press.

Parry, B. 2005. 'The New Human Tissue Bill: Categorization and Definitional Issues and Their Implications', in *Genomics, Society and Policy* 1(1): 74–85.

Peers, L. 2003. 'Strands Which Refuse to Be Braided', *Journal of Material Culture* 8(1): 75–96.

Piot, C. 1999. *Remotely Global: Village Modernity in West Africa*. Chicago: Chicago University Press.

Rabinow, P. 1996. *Making PCR: A Story of Biotechnology*. Chicago: Chicago University Press.

Scheper-Hughes, N. and M. Lock. 1987. 'The Mindful Body: A Prolegomenon to Future Work in Medical Anthropology', *Medical Anthropology Quarterly* 1: 6–41.

Scheper-Hughes, N. and L. Wacquant. 2003. *Commodifying Bodies*. Thousand Oaks, CA: Sage.

Sharp, L. 1995. 'Organ Transplantation as a Transformative Experience: Anthropological Insights into the Restructuring of the Self', *Medical Anthropology Quarterly* 9(3): 357–89.

———— 2000. 'The Commodification of the Body and Its Parts', *Annual Review of Anthropology* 29: 287–328.

———— 2001. 'Commodified Kin: Death, Mourning and Competing Claims on the Bodies of Organ Donors in the United States', *American Anthropologist* 103: 1–21.

———— 2005. 'Re-Socialising the Technocratic Body'. *Social Bodies workshop, September 2005*. Girton College, Cambridge, UK.

———— 2006. *Strange Harvest: Organ Transplants, Denatured Bodies, and the Transformed Self*. Berkeley: University of California Press.

Silverman, S. 2003. 'Forward', in A. Goodman, D. Heath and M.S. Lindee (eds), *Genetic Nature/Culture: Anthropology and Science beyond the Two-Culture Divide*. Berkeley: University of California Press, pp. ix–xiv.

Sinclair, S. 1997. *Making Doctors*. Oxford: Berg.

Strathern, M. 1981. *Kinship at the Core*. Cambridge: Cambridge University Press.

———— 1988. *The Gender of the Gift: Problems with Women and Problems with Society in Melanesia*. Berkeley: University of California Press.

Toren, C. 1998. *Mind, Materiality and History*. London and New York: Routledge.

Van Wolputte, S. 2004. 'Hang on to your Self: Of Bodies, Embodiment, and Selves', *Annual Review of Anthropology* 33: 251–69.

Vaughan, M. 1991. *Curing Their Ills: Colonial Power and African Illness*. Stanford, CA: Stanford University Press.

Viveiros de Castro, E. 1998. 'Cosmological Deixis and Amerindian Perspectivism', *Journal of the Royal Anthropological Institute* 4(3): 469–88.

Zimmerman, F. 1988. *The Jungle and the Aroma of Meats: An Ecological Theme in Hindu Medicine*. Berkeley and Los Angeles: University of California Press.

Chapter 1

AGED BODIES AND KINSHIP MATTERS: THE ETHICAL FIELD OF KIDNEY TRANSPLANT

Sharon R. Kaufman, Ann J. Russ and Janet K. Shim

> *Renal transplantation has emerged as the treatment of choice for medically suitable patients with end-stage renal disease. More than 60,000 patients await kidney transplantation and are listed on the United Network for Organ Sharing (UNOS) recipient registry. Live donor renal transplantation represents the most promising solution for closing the gap between organ supply and demand.*
> Journal of the American Medical Association, 2005

> *How, then, might one begin to mark out the specificity of our contemporary biopolitics?*
> Nikolas Rose, 'The Politics of Life Itself', 2001

Kidneys and the Social Body

What happens to sociality in the face of medical transactions that enable the transfer of organs from one body/person to another? The activities that constitute clinical life extension comprise one site for the governing of life and kinship and the emergence of new forms of social participation in which biological knowledge and identification are foregrounded. Our ethnographic example at this site is kidney transplantation for older adults, and we ask: in what ways are bodies relational – and what is at stake in those relations – when longevity *at older ages* becomes an object of intervention and apparent choice? We are concerned with how family and other relationships are implicated in a biopolitical field in which certain medical practices (along with their legitimating financial supports), *and* the desire for and expectation of a longer life, *and* changing ideas about 'normal' old age, *and* family obligation become intertwined.

The substance of the body, as ground for consideration about affective ties, stands out as a dominant cultural feature in this example, particularly for how kinship is 'done'. Indeed, the materiality of the body and its relationship to notions of health has become an important frame for ethical judgements generally (Rose 2001). We explore here the kinds of social obligations and, thus, moral order (Mauss 1967) at stake and in play when the age for transplant moves up beyond seventy and, especially, when living donors come from the succeeding generation.

Families have always been implicated in medical care and its social ramifications, but the fact that older lives are routinely extended today by medical technique infuses familial obligation with judgements about the relative worth of clinical interventions, disabling chronic illness, suffering and life itself. While obligation remains an unchanged and essential cultural fact, it is experienced now, in affluent sectors of US society, through the option or even imperative of asking and accepting, offering and giving parts of one's body (in life or in death) to another. The very routine-ness of transplant procedures extends awareness about obligation, judgement and action to the body itself and to its uses, we found, via the ever-present *potentiality* of being a donor or recipient. That potentiality, a relatively recent feature of socioclinical life around the world, gives the always already relationality of the body a relatively new form of expression.

The number of kidneys transplanted to people over the age of sixty five, both from live and cadaver donors, has increased steadily in the past two decades in the US.[1] Transplants are routine in the seventh decade of life and sometimes are performed into the early eighties. Cadaver kidneys from donors over the age of fifty are sought and are available so that they can ease the shortage of transplantable kidneys for older recipients. In addition, living kidney donation is on the rise for all age groups, exceeding cadaver donation for the first time in the US in 2001. Of the 15,135 kidneys transplanted in 2003 (among all age groups), 47 per cent were from deceased donors, 53 per cent were from living donors. Among the 1,684 kidneys transplanted to people aged sixty-five and over in 2003, 513 (nearly one-third) came from living donors, 295 of them from adult children.[2] The phenomenon of live kidney donation for older recipients opens up new dimensions of intergenerational relationships and medical responsibility as yet unexplored.

Ironically, even while biological relatedness is no longer a clinical imperative due to improved immunosuppressive techniques,[3] the nature of relatedness matters enormously, we found, in terms of one's sense of duty and responsibility for the health and longevity of a family member (or friend) with end-stage renal disease. Spouses, siblings and adult children are mostly the first to volunteer to donate when they learn of the need for medical intervention to save a life, and they represent the vast majority of living donors in the United States. Other relatives, friends and acquaintances volunteer to donate as well.

In our own observational context,[4] adult children (in their thirties, forties and fifties) are donating kidneys to their parents (currently, those who are in their sixties and seventies). Nephews and nieces, spouses, other relatives and friends are donating kidneys to older persons as well. While donation to strangers and the global traffic in illegal organ sales have been reported in the media,[5] heightening consciousness both about the desire and need to give and receive and the relative worth of life and health around the globe, there is little public knowledge or discussion of the transfer of organs from younger to older people. Yet the latter phenomenon is affecting more and more families (Grady 2001; Dowd 2003). In addition, the broadening of the eligibility criteria for legal kidney donation over the past three decades – from healthy, relatively young cadavers and living family members to the inclusion of older cadavers with diseases and non-related living individuals – marks a change in the credit/debt dimension of biomedical subject-making. Along with this expansion in prospective and actual donor type, the recent, highly visible cases of apparently purely altruistic, philanthropic living kidney donation to strangers provide evidence of an expansion of the claims of citizens on one another.

Framing an Ethic of Care

At the site where old age, end-stage disease, clinical medicine and the contemporary, ubiquitous expectation for increased longevity meet (in the US), we are aware of a new kind of ethical knowledge that is emerging through routine clinical treatment. We refer here not to bioethics and the well-known parameters of that still largely philosophically based discipline for clinical decision making in individual situations, but instead to a diffuse ethical field,[6] located throughout contemporary social life, that contains certain forms of reasoning, expectation and judgement about medical possibility. This ethical field is *characterized by the difficulty, sometimes the perceived impossibility, of saying 'no' – even in late life – to life-extending interventions,* and it is constituted by the following three features. First, the purported choices clinical medicine now provides to patients, prospective patients, their families and their doctors, regarding whether and when to employ life-extending procedures and whether and when to stop them, are not really choices at all. Rather, choice elides into routine treatment. When procedures are shown to be effective at ever-older ages, they become routine and thus expected and desired by clinicians, patients and families (Koenig 1988). When techniques become less invasive and associated with lower mortality risk, consumer demand for them and ethical pressure to make them available both increase. Finally, procedures that are seen to be relatively low risk quickly become standard practice. Treatment rationales and patterns are thus already scripted for patients and doctors, well before any actual 'decision-making' begins.

Second, the availability of interventions as therapeutic possibilities elicits hope and expectations for cure, restoration, enhancement, and improved quality of life. In turn, the boundaries between medicine's focus on cure, prevention and life prolongation are increasingly blurred in the desire to maximize life. These two features support and give form to the third: the nature of caregiving and love has expanded so that expressions of care (both medical and familial), affection and value are explicitly tied to clinical acts that either extend life in advanced age or allow 'letting go'.

The broadening of medicine's scope over the management of old age as well as illness per se (Arney and Bergen 1984; Kaufman 1994) reinforces and sustains these three features of the ethical field which are mostly background to the daily activities of health care practitioners and patients in the US. Yet, they are the basis through which we come to understand the value of medical care and the uses of the body, our connections to one another, and the social worth of the oldest lives in our society. Each practitioner, patient and family must work out an ethic of care in the course of unfolding events, active and tacit decision making, and reflection. That ethic is rarely premeditated, autonomous or deliberate. Rather, it emerges during social interaction and the unfolding of medical and interpersonal events. It is shaped by the structural features of medical specialties and the American health care delivery system, the ways in which cultural understandings of kinship and community and epistemologies of clinical medicine influence one another (Franklin and McKinnon 2001: 9), and, in the case of kidney transplantation, by the apparent mechanical ease with which body parts can be transferred from one person to another.

Expectations for ever-longer lives join expectations that routine medical treatment can extend a good life. For many, it is simply unacceptable to die at seventy-one or eighty-one when one can employ the tools of the clinic to restore health and stave off death. While access to life-sustaining treatment, at every age, remains unequally delivered in the US, the treatments themselves exist, and almost everyone knows this. Things can be done, and the family is involved. Love is actualized, often, through the commitment to a longer life and by doing things to prolong life.

The possibility of *giving away* a part of the body becomes implicated both in the demonstration of care and love and in the ability and responsibility to prolong another life. The possibility of *receiving* the body part of another – the always already quality of this social fact – becomes part of the calculus by which the potential risk to another life and the sacrifice of another's bodily integrity are weighed in relation to the value of extending one's own life and improving one's own well-being. The expansion of love and care, and of expectations about medical intervention to include the possibility of the transfer of bodily substance, are cultural facts now regardless of whether the acts of giving and receiving actually occur. To be confronted with end-stage kidney disease (of self or other) in a context in which transplantation and live donation are normalized is to existentially engage the sociality of the body.

In this way, living donor transplantation joins the new reproductive technologies (Franklin 1997; Rapp 1999; Thompson 2001) in broadening the field of ethical action to include proactive deliberation about the uses of one's own body and the uses of the bodies of others to achieve health and promote new or extended life. In the case of organ sharing, biology is not mobilized to configure kinship (Thompson 2001: 175, following Schneider 1980), but rather, biology, as corporeal substance in the form of 'one's own body' (in the US at least) becomes the central object in articulating what is 'natural' about the demands of interpersonal relationships – and thus what one can ethically claim from others – and in deciding what to do.

Intersubjectivity and the Ethical Field

Intersubjectivity is the quality, the pre-condition, that makes possible this ethical field. By 'intersubjectivity' we refer to the inextricable joining of self-knowledge, interpersonal relations and social/ethical participation that constitutes, in large part today, the making of subjects. We might say that, together, embodiment and social emplacement shape experiences of self and other, the parameters of identity and subjectivity and the nature of our relationships.[7] Contemporary biomedicalization (Clarke et al. 2003) inflects bodily experience, self-making and sociality with particular kinds of knowledge: about how the body is implicated in the ethical; for what becomes naturalized about the human, the family and the life span; and for objectivizing the subject (Foucault 1983: 208). In the case of the life-extending medical practice we describe here – kidney transplantation in later life – the body, *understood as transferable parts*, becomes a central feature of intersubjective knowledge and relations. Organs can be variously 'owned' and symbolically characterized, and they can be understood as things, persons, self and other, depending on one's point of view (Sharp 1995, 2001). There is slippage among these categories and the breakdown of body and identity boundaries creates some angst in the transplant world that has been described in detail (for example, Fox and Swazey 1992; Lock 1996, 2002; Sharp 1995, 2001). Yet the 'good' of transferability (as illustrated in the opening epigraph) is widely accepted and is evident in the growing demand for organs. That demand emerges from the broad expectation that end-stage kidney disease need not be fatal and the fact that kidney transplantation is a standard medical treatment for end-stage disease. There is thus growing opportunity for family (and other) relationships to be ethically circumscribed via biological exchange.

Life-extending medical practices in general and kidney transplantation in particular present useful case studies for intersubjectivity for two reasons. First, intersubjectivity includes somatic experience. Being-in-the-world is conditioned by the body-self and thus commitments, intimacies and a naturalized, embodied (and thus non-reflexive) ethic flow from that experience (Merleau-Ponty 1962; Csordas 1990, 1994; Kleinman 1997). Second, participation in standard-of-care life-extending medical treatments involves – indeed, cannot be separated from – the formation of self-responsibility and rights on the one hand and social connection and obligation on the other. These two aspects of subject-making are always co-constitutive. They are informed, more and more often, by clinical developments and imperatives.

On the self-responsibility side of the equation, older persons (and their families) in the US come to understand their bodies, lives, possibilities and futures – including what constitutes the 'normal' life span – in terms of their options about medical interventions that may or do extend life by postponing, altering or ameliorating an 'end-stage' condition. Nikolas Rose notes that by the late twentieth century, knowledge of the living body 'became instrinsicially linked to interventions that transformed those living bodies' (Rose 2001: 14). Life, health, illness and death became objects to be acted on via the instrumental techniques that the biological sciences and clinical medicine offered (see also, Franklin 2000; Rabinow 1996, 2000). One's own biological destiny, and that of one's progeny, are no longer fixed, immutable. Prevention, enhancement and intervention are possible, even into advanced age. We can choose to extend late life via medical action because biomedical technique has extended choice to every aspect of existence (Rose 2001: 22), including the timing of death. We can choose not to intervene as well. Either choice is made within a cultural framework of increasing normalization of life-extending medical procedures.

On the other side of the equation, the nature of intimate family relations, intergenerational commitment and one's obligation to loved ones and to strangers in need, for instance, are affected more and more by biological knowledge and medical imperatives. Not only is the gaze of biomedical science and clinical medicine shaping the ways we understand the self, the old body, aging and our interpersonal commitments but, also, the multiple (and even contradictory) ways we understand aging, old age and family matters are changing clinical practices and the sense of responsibility towards the self and others that underlies those practices (Estes and Binney 1989; Kaufman, Shim and Russ 2004). Clinical solutions and management strategies for aging bodies shape family and social dynamics and create new manifestations of discursive medico-political power. Our claims on ourselves and on one another are both visible and questionable in an unprecedented way: through the promises of the body enabled by an only partially articulated contract – about how bodily substance may be invoked in ethical action – among the clinic, the patient and the donor.

In the case of kidney transplantation, others are deeply implicated in sick patients' self-responsibility to remain alive and to be as healthy as possible. Physicians and family, in their press to treat and desire and duty to 'make live' (Foucault 1980), create an opening for a dual obligation for older persons – to be treated and to stay alive (not just to get well, as Parsons (1951, 1975) noted). That obligation, if taken up by the patient/consumer, is a pledge (as well as a burden) *to others* that one will follow the logic of medical intervention wherever it leads. Kidney transplantation at older ages is becoming ordinary, unremarkable (like the normalization of high-tech reproduction and many other medical procedures) and so reflects and supports the unavoidable connection between judgements of value and biological existence and between health and social worth (Rose and Novas 2005). Its growing normalization (despite the fact that some do reject the procedure) highlights, in addition, how 'we make the ties that bind' (Laqueur 2000) between and among family members, generations and strangers, according to the contingent intersections of biomedical expertise, the ubiquitous demands of consumer society on the self, and the culture of the contemporary health care delivery system. Even if *potential* donors and recipients decline to use their bodies in this way – that is, do not become *actual* donors and recipients – transplant medicine has created that potentiality, and decision making for everyone involved responds to that possibility. Our goal is to outline how the clinic, patient and family, together, shape a particular kind of bond between biological identity and human worth and a demand for an old age marked by somatic pliability, renewability and a claim of responsibility that merges and organizes the (bodily) obligations of self and other.

The Obligation of the Gift

Mauss's (1967) explication of the 'symmetrical and reciprocal' obligations of gift exchange – to offer and give, to receive and accept and to repay – was effectively taken up by Renee Fox and Judith Swazey (1992, 2002) in their nearly forty-year documentation of the impacts of organ transplantation on patients, families, medical practice and American society. The 'tyranny of the gift' they describe – the imperative to offer and give, accept and receive the gift of an organ and the gift of a longer life, regardless of health or suffering, guilt or desire, and the painful 'creditor-debtor vise' (Fox 1996: 254; Siminoff and Chillag 1999) that may envelope givers, receivers and families – has additional social ramifications, it seems, when the direction of organ transfer is from younger to older persons. We heard from some recipients and prospective recipients, for example, that this direction of transfer is 'unnatural'. Some health professionals noted that the passage of organs from younger to older persons is inappropriate from the standpoint of medical goals and use of resources. Yet most patients with kidney disease want treatments that will enable them to live – and to feel as good as possible. We found, in addition, that many recipients feel obligated to live for their families, and donors feel duty bound to allow their parent (or older relative or friend) to continue living – and to facilitate that continued life.

Two socio-clinical facts form the basis of this obligation. A transplant from a living donor has a comparatively high success rate and a better prognosis than the continuation of renal dialysis (Wolfe et al. 1999; Chkhotua et al. 2003; Mandal et al. 2003). And, insatiable demand for transplants from older persons with kidney disease arises from and supports the growing normalization and naturalization of older kidney recipients. These realities put enormous pressure on family members, especially adult children, to donate (Fox and Swazey 1992, 2002), whether or not health professionals specifically urge, suggest or opine against living donation. Not only must older persons continue to live, but also younger persons must give. Not only are 'routine' treatment patterns scripted, but also old and young are deeply committed to one another, sometimes beyond deliberative choice. 'It's just something you do, no question about it', we were told repeatedly by donors. Many prospective and actual donors view donation as simply 'giving back' for all a parent or other relative has done for them.

The gift is 'not a big deal', we were often told by donors after the fact, because medical technique has made kidney donation ordinary, easy and of negligible risk to the donor. The gift is simply the best way to extend one life while, according to the medical discourse in the US context, not diminishing or risking another (Davis 2004; Davis and Delmonico 2005).[8] Yet it is, fundamentally, a sacrifice of the wholeness of the body and a non-reciprocal bargain. Among recipients there is much more ambivalence about it, as we will show. Often, recipients' desire to refuse the gift is muted, masked and overwhelmed by the routine-ness of accepting, the discomfort or refusal of dialysis, increasing serious illness, the willingness, enthusiasm and persuasion of a donor and the medical truth that a living donor will provide the best health outcome. These are powerful influences on what happens in the clinic.

Physicians shape the options for older patients and their families and encourage living donation when they stress the time-sensitivity of the decision: persons over seventy may become 'ineligible' for a transplant if they wait too long, that is, if they remain on the four- to six-year UNOS waiting list (at the time of this writing) for a cadaver organ. The older one is, the more precarious one's health may become in a few years, and thus the greater urgency, both in order to be medically suitable for transplant surgery and for transplant to be successful. The waiting time for a kidney from a cadaver donor over the age of fifty or sixty is not as long as the wait for a 'younger' kidney, and there is virtually no waiting time for a kidney from a living donor. Moreover there is little, if any, increased health risk, doctors point out, to donating and then living with one kidney, even for decades (Davis 2004; Davis and Delmonico 2005). Patients and families quickly learn that their choice must take into consideration time and age. For instance (physician to a 77-year-old man with heart disease): 'Realistically, you'll have to have someone donate you a kidney if you have a chance of getting one'; (physician to a 71-year-old woman): 'I think getting you a kidney would be a great thing. But the sooner the better. It could be 5–6 years if you wait for a cadaver donor, especially because of your blood type'.

Though health professionals we observed always indicate to patients that a kidney transplant carries some risk and that living with a transplant will not eliminate all health problems or the lifelong need for medication, they do stress that a transplant will free the person – as nothing else can – from the physical side-effects and functional limitations of dialysis. If the patient is considered a good candidate for transplant by physicians, regardless of age, life extension and better quality of life are available (Wolfe et al. 1999). Medicare covers a great deal of the costs of transplant medicine; Medicaid covers some costs as well. This is not a procedure only for the affluent.[9] In large measure, then, if the patient has financial and other access to transplantation medicine, it is up to the patient, and this fact marks a new geronto-ethics.

> Physician: It depends on how active you are and want to be. Transplant frees you … People who want to be active want the added years. It all depends on what you want. It's a personal choice.

> Patient, age 71: Now that dialysis has started, I feel I'm in a holding pattern, a waiting pen. Before I started, I didn't realize I would think of it this way. I could never stay on it for years. I see transplant as my liberator, the light at the end of the tunnel.

Thus one choice that is often presented to older patients and their families, or that emerges for them during the clinical encounter, is either, first, a longer or shorter waiting period for a cadaver kidney – and possible death or greater infirmity while waiting – or, second, a living donor, 'someone you know'. Starting or remaining on dialysis is always an option, but health outcomes are better with a transplant (Wolfe et al. 1999; Segev et al. 2005) and so, when health professionals consider patients good candidates for a transplant, they encourage them to proceed. Also, in our experience, the vast majority of patients do not want to remain on dialysis if they can receive an organ. The shadow of death, the last and often unstated choice in these conversations, along with the negligible risk to living donors, hangs over the question of what to do.

The cultural context in which transplantation is routine and living donation is always already a possibility, coupled with growing numbers of living kidney donors for older recipients (Mandal et al. 2003), adds a new dimension to kin obligation by forging a powerful, unprecedented connection between one's sense of uses of the body and one's idea of duty. Biomedical knowledge about the success of transplantation at older ages contributes to a dual truth about older bodies: they are a terrain for improvement and life extension and an objectification of the way family members, friends, acquaintances and strangers may belong to one another (Edwards and Strathern 2000). The general success of kidney transplantation in later life opens a realm of possibility about future health and the self by telling us who we can be as recipients and donors (regardless of whether we choose, as individuals, to give or receive). That success both describes and paves the way towards a new manifestation of commitment (Franklin 2001; Kaufman, Shim and Russ 2004). We suggest that the obligations surrounding gift giving and gift receiving, combined with medical technique, are contributing to a sense of a 'natural' act, a 'natural', longer life span and a 'natural' expression of love and belonging.

The site of ethical judgement and activism about longevity and mortality, in the case of kidney transplantation, is one's regard for the body of another, especially one's assessment of the role of the gift in that relationship. This decidedly corporeal feature of ethical relations is inextricably joined to the widespread quest for greater longevity and enhanced 'quality of life' that are assumed to be available. Thus the difficulty of saying no to life extension joins the tyranny of the gift in an unprecedented way.

Transplant Subjects

We are interested in mapping a phenomenology and practice of responsibility and obligation in this ethical field, though we are well aware that our small-scale ethnography, to date, can only partially expose its characteristics.[10] For kidney recipients, prospective recipients, donors and prospective donors, we tracked and elicited the following features of self-making and social connectedness: (1) whether one will accept a living donor kidney, and, if yes, where one draws the line for acceptance – spouse and siblings? Children? Co-workers? Friends? Strangers?; (2) the time it takes to offer, give or accept a live donation, or be persuaded to accept; (3) the degree of aggressiveness in the search for a live donor; (4) the degree of willingness to begin renal dialysis at all, stay on dialysis for an unknown duration, or use dialysis as a short-term stop-gap measure; (5) the degree of urgency about asking, accepting and offering; (6) the order of things: seeking a live donor first (and among whom) and viewing cadaver donation as a last resort, versus deciding to wait for a cadaver organ first and turning to a live donor only when one's health deteriorates significantly; and (7) the worth and value of one's own life, relative age and relative need vis-à-vis another. Each of these features that marks the transplant experience contributes to somatic self-fashioning and the biologization of intersubjectivity (Biehl, Coutinho and Outeiro 2001: 116). Moreover, these features both reflect and serve the clinical contribution to a certain kind of intersubjectivity – older persons *in need* of and *eligible* for transplants and their willing and obligated, potential and actual donors. That need and eligibility form the basis, the frame, for the work that prospective donors and recipients do when they engage the prospect of transplantation.

Although family relationships are central to kidney transplantation today, our conceptual aim here differs from explorations of what counts as kin relatedness that characterize the 'new kinship studies', especially those that analyse the social practices surrounding the new reproductive technologies (Franklin and McKinnon 2001; Carsten 2004). That is, potential kidney recipients and donors do not *elaborate* kin categories as do those engaged with reproductive medicine, and thus our goal is not to focus on the processes of conversion between the domains of the biological and social (Carsten 2001: 50), or how people deliberate what counts as a *biological* tie (Franklin 2001: 302), or how certain relationships are naturalized and made less ambiguous in the process of medical intervention (Thompson 2001: 175).

Rather, our aim is to show how the social fact of the life-saving nature of living organ transfer is always already present when potential donors and potential recipients, along with others in their social world, express their moral claims on one another. We found that while those claims certainly *invoke* relatedness, especially the emotional ties that accompany biological kin relations and those deemed to be 'like' kin, our informants do not, specifically, elaborate kin categories per se in order to explain their participation in these procedures. It is precisely the 'naturalness' of the kin (or 'like-kin') relationship – and thus the absence of needed elaboration – that justifies and normalizes consideration of the gift, whether one offers and gives ('It was just the thing to do, to donate – no question about it'; 'This is the way it should be') or not ('I will never accept a kidney from a living person; I would rather die than accept that responsibility').

Our theme is the *consequences* of the normalization of kidney transplantation as a clinical/cultural practice. Claims are made and ethical acts are called for because of the existence and significance of ties – whether biologically and/or socially defined – which are deemed to be in the service of a life-saving act. Those claims resonate in new ways today because the infirmities of 'old age', including end-stage kidney disease, are now often treatable. In this example claims and then choices derive from standard-of-care medical practice. The biomedical truth about the social worth and clinical efficacy of kidney transplantation is seen to operate as a cultural system (following Schneider 1980; Strathern 1992; Franklin 2001) shaping relationships in the form of an ethic of care. The work done by all players involved is to rationalize and emotionally come to terms with the prospects of offering and giving, accepting and receiving, and, importantly, deciding where one draws the line in networks of family (however defined), friends and strangers and across relationships that are marked, in the case of relatedness, by protection, love, obligation and indebtedness and, in the case of strangers or casual acquaintances, by apparent altruism and its acceptance.

The potentiality of offering, giving, accepting and receiving a kidney puts pressure on parent-child, spousal and other meaningful bonds. Within those bonds, offering is a sign of unconditional love, sacrifice and the strength of the bond. For adult children, offering is a sign of thanks and payback for all the protection and nurturance a parent has given over a lifetime. Between spouses, offering reflects mutual care, support, sacrifice and its acceptance. For recipients, taking the gift is the sign of thanks for love, mutual obligation, and acknowledgement that the gift is appropriate.

A wealth of studies in science, technology and medicine over the past decade illustrates the ways in which clinical truths/knowledge have profound interpersonal and intersubjective effects. In the case discussed here, clinical knowledge leads to an ordering within the social worlds of both potential recipients and potential donors, of what kinds of relationships, if any, will be recognized as socially and emotionally eligible and appropriate for organ transfer. In the aggregate, kinship bonds rank the highest as we write this in 2005. But offers to give from friends, acquaintances, co-workers and, most recently, from strangers, and acceptance of those offers, appears to be a growing phenomenon in the US for adult kidney transplant (Nolan et al. 2004). The symbolic space (Carsten 2004: 181) that kinship (however described) occupies in living organ donation may, in fact, be shrinking.

Recipients

The recipients and prospective recipients we spoke with articulated a broad range of ethical opinion about the need and urgency to ask for and accept an organ, the obligation not to ask and not to take, and the responsibility either to wait years for a cadaver donor or to quickly solicit one's own potential living donors. We observed a range of engagement styles with transplant medicine, from proactive self-education about transplantation, discussion with physicians and strategic avoidance of dialysis to passive acceptance of whatever treatment a physician recommended. Neither ethnicity, immigrant status nor gender determined the ethics and practice of seeking or accepting an organ among our small but diverse sample of thirty-three kidney recipients and twenty-eight prospective recipient interviewees.

Patients articulated the ways in which they put into practice an ethic that balanced self-responsibility for an extended life with family obligation and the pursuit of health with selfishness and irresponsibility. Mostly – though not exclusively, as we illustrate below – the difficulty of saying 'no' dominated the ethical field and charted the terms of engagement. Patients expressed a slippage between the expectation of life extension and the obligation to undertake it – they knew longer life was possible via medical intervention, but they did not always know, immediately, how to ethically rank their options for achieving longer life.

Some people want a kidney transplant as soon as (or before) they learn they have 'end-stage disease' and they line up prospective donors before their first visit to the transplant evaluation clinic, before talking with health professionals about transplantation. Some of them organize a list of potential donors in order to avoid starting dialysis at all, and regardless of whether their physicians suggest or offer transplantation as a treatment. For others, the possibility of a transplant (live donor or cadaver) dawns slowly, over a period of weeks, months or years after dialysis begins. Those in that latter group come to the transplant clinic for a work-up and evaluation, not necessarily because they feel ill or seek a transplant, but rather because their primary care doctor or renal specialist suggests they do so. They are put on the waiting list for a cadaver donor. At that time, they may (but may not) consider a living donor even though some of them are ambivalent about a transplant.

Even the most passive, ambivalent patients – if they decide to remain on the waiting list – can receive a transplant simply by waiting for one (provided they continue to be medically eligible) because the clinical pathway will move in that direction. In those cases, it takes more psycho-cultural work to reject a cadaveric kidney than to accept one because medical interventions to prolong life are routine, kidneys provide better health outcomes than long-term dialysis, patients want to live, and everyone around the patient (family, friends, health professionals) supports transplantation as the appropriate, best intervention.

People who actually make decisions of some kind in regard to seeking or accepting a kidney from a living donor express a range of ethical imperatives. From those who refused to accept a live donation we heard, for example:

> The only thing they asked me was, did I have anyone in the family who was willing to donate a kidney for the transplant. And I flat out told them, no. My nephew was willing to donate, and I said, no way, I wouldn't do it … If his other kidney failed, he would have been in trouble, and I couldn't give him back the kidney. I even asked that question to others: Can I pass this kidney on after I pass on? Can they re-use it for somebody else? And they said no … I morally would not accept a live donor. It is not worth saving my own life to take on that moral burden. I waited for a cadaver kidney and I waited five years.

> I do not want a living donor. If something happened, I wouldn't like it, whether it's a relative or not. Let's face it. I'm not that young. I only have so much to go. I don't want to enjoy it at someone else's expense. I don't want to put someone else in jeopardy. I've got to go sometime. If I've got to go, I've got to go.

Those who refused a living donation could not reconcile the potential risks of death or debility to a donor with the gift of a longer life for themselves, regardless of their relatedness to that prospective donor. In the shared calculus of these patients, the prospect of inflicting bodily harm on another loomed larger and weighed heavier than the desire or need to extend their own lives, and each took his/her chances on waiting for a cadaveric donation. Their judgements about the worth of their extended lives relative to the potentially compromised life of another was unequivocal, and they did not categorize the acceptability of the gift by ranking or defining kin and non-kin relationships.

In contrast, from individuals who actively sought donors, we heard:

> In the beginning, my doctors did not offer or discuss transplant. When I learned, two years later, that I wouldn't have to wait around on dialysis if I had a live donor, my wife lined up fourteen people who would donate. One of the transplant nurses told me that that was nothing. She's seen people who have lined up a hundred prospective donors. You have to be proactive. You can't just sit around, or you'll die waiting.

> When the team said, 'We'll put you on the list for a cadaver transplant,' my granddaughter brought up live donors and said, 'We're ready.' We had the forms filled out by seven people in 24 hours. I didn't consider anything else – I never went on dialysis.

For these patients the naturalness and good of transplantation as standard medical treatment muted or made irrelevant any deliberation about their own relative worthiness and what they understood as the relatively minimal potential negative health consequence to a donor. Though talk of sacrifice or risk to the donor sometimes emerged as expression of a patient's moral calculus, those factors did not tip the balance away from a proactive search and then acceptance of the gift. Moreover, these patients had a great deal of family support in mobilizing potential donors from within and beyond kinship networks. Solicitation of a living organ, regardless of how widely the patient casts the net or how the donor is categorized by the patient, occurs when patients, along with their support systems (which includes health care professionals), understand transplantation first and foremost as a right and, along with donation, as ordinary medical practice.

There are intermediate ethical positions as well, between outright refusal to accept a living donor organ and proactive solicitation of one. The ethical criteria for accepting a kidney from a living donor, and then who will be considered, are idiosyncratic and often change as the patient gets sicker, adjusts to the routine of dialysis, or receives an unanticipated offer of a kidney. Some patients will not ask for a kidney but will accept one offered by someone they know. They draw a line at the place where unprompted, insistent generosity meets a request that could seem coercive because of the significance of the bond. For them, an unsolicited and insistent offer of donation is evidence enough of genuine altruism, and they thankfully accept. Insistence and persistence on the part of the donor seems crucial to recipient judgements. Their own relative passivity in the situation allows their escape from a degree of moral culpability that could, perhaps, lead them to say 'no' to the gift. Full moral agency is placed on the donor instead.

People who will accept an organ from a living person ultimately establish a hierarchy of potential donors based on any combination of the following characteristics: biological relatedness, the strength and length of tie between them and the prospective gift giver, the degree of enthusiasm or hesitation that person expresses about donation, and whether that prospective donor has responsibilities for other lives. That hierarchy reflects the construction of the ideal biosocial candidate for donation. For example: The son is running a business; he has four employees who rely on him. The daughter has two young children. But another daughter is not married, does not have children and does not have the financial responsibilities of the son. She is the ideal candidate. Or, the daughter with two children believes she is the most physically fit member of the family. She offers and is adamant about it. Or, the children have their lives ahead of them. I wouldn't accept from them because they may need their kidneys later on. But I will take a kidney from someone else, someone not as young as my children.

We often heard, at first, a negative injunction against asking one's spouse, children, siblings or other family to donate – 'I'll never take an organ from one of my children'. That negative stance reflected resignation to and acceptance of the fact that the end of life – a long life – was near and, maybe, was appropriate. 'My wife wants to be my donor, but I don't want to take from her. We've been married fifty years. I don't want to bring her into this. This is end-stage'. The daughter of a 75-year-old patient said: 'He has a sister and me to donate, but he's not accepting it. He's very close to his sister and doesn't want anything to hurt her … He says he's at the end of his age. But I say, No, that's not true, you're not at the end of your life'.

The initial refusal to consider one's family member as a donor frequently gives way to acceptance in weeks or months because some patients feel extremely ill on dialysis and/or because they want the freedom and better health that a transplant promises. Thus, sometimes, pragmatism and comfort trump principle, especially in the face of the persistent generosity of a prospective donor: 'I was getting sicker and she kept offering'. Or, 'She was saying, "Come on, Dad. My kidney isn't getting any younger".'

Pressure on patients to accept a living donation is applied from prospective donors and the structure of health care financing which makes surgery and medications affordable for large numbers.[11] While personal choice is considered by health professionals to be paramount for patients in deciding whether to have a transplant in the first place, there is relatively less choice about accepting a living donation if someone unequivocally offers because, together, enthusiastic prospective donors and the transplant team support the ethical field. Standard medical treatment that can and does treat aging and thwart dying, prospects of bodily freedom, better health and quality of life, and the fact that love and care are expressed through clinical intervention work in concert as compelling reasons to offer and accept. In the example of living kidney donation, the family (and friends, acquaintances and most recently, strangers) serve as the agent of medicine and its goals (Biehl 2004).

Moreover, the responsibility to pursue greater health and longer life bleeds into the obligations we have for one another. The following form of reasoning stood out among those patients who ultimately accepted a kidney from a spouse or adult child: my family needs and wants me to live because it is possible for me to do so, and I want to live. Therefore I need to live, so they (or some of them) will offer to donate a kidney for me, and (though it may not seem right) I must accept it. For example, one recipient recalled his clinic evaluation:

> We were sitting there and the transplant coordinator asked, 'Is there anybody in the family?' And I said, no. And my wife said she wants to be the donor. And I said, no, I don't want you to be a donor. That's way too much to ask of anybody. And they said, well, you don't really have anything to say about it. If she wants to be a donor, she can. And they did a workup and she was a match, very very close … And then as things got worse and there was no possibility of being on the [cadaver waiting] list, I guess I kind of caved in and said, well, if the match is that good, and it will work, let's do that. I could have adamantly refused, but I didn't.

> The children talked me into it. I said, I'm not taking my daughter's kidney! But other family members persuaded me. You know, I kind of went along with my older daughter's insistence, and we didn't say too much one way or another, whether I wanted to or not. I didn't want to take an organ from my child. If it were the other way around, I would have gladly given my kidney to one of them, but because it was coming, as a hand-me-up sort of thing, I thought about it a lot. It didn't feel like it was the right thing to do. Help should go the other way, from parent to child. There were periods of time I just really didn't want to do it. There was no real point in time where I decided I wanted to have it done. I was going along for the ride because things were being arranged for me.

The opportunity to continue living despite end-stage disease is handed to older patients (and their families) by contemporary clinical medicine.

There is widespread and growing expectation that one can grow older – and that one can strive to grow older, despite chronic disease or even terminal disease – without feeling it, without an embodied sense of aging (Katz and Marshall 2003; President's Council 2003). 'I wanted to get back to a normal life. Dialysis was just not acceptable. We are too active', said a 76-year-old man who rather easily accepted his daughter's kidney.

Overall, one can summarize recipients' ethic of care around two themes. First, recipient 'choice', we found, is most clearly expressed in instances of refusal of a living *but not a cadaver* donation. Those who refuse living donation do not exclusively refuse offers from family members and do not point, ultimately, to the bonds of kin relations as the reason for that refusal. Rather, they name unacceptable risk to all potential live donors and/or their own relative worthiness in terms of age and nearness of the end of a 'natural' life as reasons to say no. Second, refusal often gives way to acceptance as health deteriorates or as donors persist in offering, because the stakes of life and relative health, the encouragement and guidance of the health care team and family, and the routine success of kidney transplantation, together, act as imperatives to go ahead with live donation, *regardless* of the stance of the recipient.

Donors

Donors' and prospective donors' side of the story of obligation, gift giving, choice and no choice reflects their understandings of urgency and necessity in the life-and-death matter of kidney disease and its treatments. Their understandings include the expectation that mortality can and should be pushed back into the hazy future. Thus, donors and prospective donors express a shared ethical clarity about their responsibility. From a daughter: 'There was no choice, no decision-making. This was simply the thing to do – to donate a kidney. He needed one. I could save his life'. From a friend: 'I don't think of it as a great thing. I just think of it as a normal thing that people you know would do ... I just think it's a natural thing'.

We heard many variations on this theme, including the widely shared assessment that long-term dialysis was not a viable option: dialysis is hopeless, a slow death, causes problems, restricts life; donating a kidney is restorative, life giving, freeing. Some donors came to this realization early in their parent's, spouse's or friend's kidney disease. Others watched the deleterious effects of dialysis on the patient, for months or years, and then realized they could donate and so offered to give – or then offered more insistently until the patient said yes.[12]

Ethical clarity about the obligation to donate emerges from and is intimately linked to a sense of family solidarity, and adult children donors expressed a connection among family solidarity, the ethics of the situation and their sense of self: 'We're full-blooded Greek'; 'We're Italian/Spanish'; 'We're Dutch'; 'We're Japanese-American'; 'so you stick by one another. Family is the most important thing, and this was a matter of the family unit'. The ultimate importance of survival of the family unit was poignantly expressed in the unusual case of two sisters each donating a kidney, eight years apart, to their father:

> When we came here from Denmark in the Sixties, it was just the four of us. My sister and I were 2 and 4 years old. My dad had $25 in his pocket, and no job. That was the only time he was really scared. And he built his life, and we're a very close-knit family ... This journey is not about choice. It's just something that you do. We weren't forced to ... We were willing to sacrifice our lives to maintain the integrity of the family unit.

Family suffering and perseverance were other themes in stories of donor and prospective donor obligation and solidarity. An adult child's offer and gift is understood in relation to what the family has been through already. Expressions of family solidarity, what you do for another family member, were not going to stop when the parent became older and sick with kidney disease. Live donation allows family solidarity to be expressed through the gift of bodily substance. The strength and compelling nature of a tie to a parent can be enacted in the transfer of a part of the body across generations, thus lending support and longevity to the entire family.

Among our donor and prospective donor informants there was a large range in the timing of offers to give that resulted from when, during the patient's illness and medical work-up, the donor realized she/he could, and should, make the offer. That timing is influenced by proactive patients, as illustrated above, but also by the independent dual realization, among family and friends, that they can offer and that their relationship to the patient prompts their willingness and/or obligation to do so. That realization may or may not be born of the urgency of the life-and-death stakes of the situation.

One daughter said,

> We were watching him die. I could prevent his death. I wanted him around many more years. It was, in part, self-interest. I offered immediately, no doubts whatsoever in my mind about what to do. I volunteered before he, or anyone, raised the subject at all. I never was burdened or bothered by a sense of risk. There was no question in my mind.

Another daughter explained,

> When they told my mother she would have to start dialysis I volunteered. They never mentioned it [live donation] at the clinic when she went for her evaluation. I could save her life, and she wouldn't have to go on dialysis. My immediate reaction was to donate. My mother was, frankly, relieved.

> The moment he started dialysis, I offered. –'Dad, you need a kidney.' It was the obvious thing to do. You don't have something that your body needs to fully function, and it's quite possible that I might have something that you can use. I'm the youngest and I think it was – 'Nobody is cutting into my baby girl' but after nine months getting beaten up on dialysis, one day Dad said, 'This is getting old.' I said, 'Are you ready to get me tested?' And he said, 'Yes, I think I'm ready'.

Future health risk to the self never weighed heavily (if it was considered at all) on the donors we interviewed.

Like recipients, donors described how a hierarchy of donation within the family was established before medical testing for an adequate match begins. An ethic of appropriateness has to be worked out:

> All of us [siblings] wanted to give to dad. We argued among ourselves about who should be tested first, who should get to donate. It was always a matter of which one of us gets to donate, not if we should donate.

A daughter who donated to her father said,

> We all have different roles in the family. My brother's married and has two young children. And I think if the chips had really come down, he might have stepped forward. My sister is married. I'm single. I didn't have to consult anyone. So that's just the way it was. And so many people said, 'Are your brother and sister getting tested?' And it was kind of like, well, I worked. I worked out. Why bother?

A 75-year-old man who donated to his 70-year-old wife said,

> Naturally the kids said, 'Take me.' They automatically offered. But we didn't want the kids. They're young, and they may need their kidney later on. We both felt that if I was compatible, I would do it. And they were very happy with that. I was really worried, during the testing, that they would find something, or I would be too old, or I would have an abnormality in the other kidney, and I wouldn't be able to give. And it didn't happen and we were very pleased. And I'm sure the kids were relieved.

Personal health, age, time available, responsibility for others beside the prospective recipient and sheer insistence all figure into the calculus of potential donors' own rankings among themselves. While donors offer and give 'without question', self-judgement about giving is intimately tied to perceptions of how one will be judged within the family, one's susceptibility to guilt in this regard, and the ways that the prospect, hope, ease and routinization of transplantation shape the value of (extended) life and tie it to social, especially familial, worth. Obligation and connectedness to others are mutually shaped by the ways in which reciprocity, duty, indebtedness and love are enacted in giving and receiving a part of the body.

The Shape of Freedom

Enacting one's own freedom – through one's right, obligation and commitment to health and long life – is a complex and demanding enterprise. Growing older without aging (Katz and Marshall 2003) is the form that freedom takes in our era of routine life extension, an era that might well be characterized by the merging of the 'right to live' and 'making live' (Rabinow and Rose 2003). The case of kidney transplantation offers one empirical example of how that freedom is understood by prospective and actual donors, recipients, families and health professionals and how it is organized in and by an aging society in which the biomedicalization of life is a dominant form of discursive power.

Death and choice about death are now thoroughly medicalized in the affluent sectors of the West, and the biopolitics of the 'right to die' have been clearly (though narrowly) articulated in the last decades – control over the timing and means of death, symptom management, humane rather than heroic intervention. Subjectification vis-à-vis death is marked by the right, responsibility and freedom (if one is fortunate enough) to authorize the location, style and timing of one's own end (Kaufman 2005). By comparison, while life itself and the stages of life (including the foetal) are also thoroughly medicalized, the freedom to 'make live' in late life opens up a new truth about the relationship among intersubjectivity, obligation of and through the body, and the desire for extended life. That truth is constituted, first, by an ethical field in which life prolongation by medical means is a routine option and the pathway to transplantation is wanted and encouraged; second, by the work done within that field to interpret one's own personal ethic of care regarding offering and giving, accepting and receiving; and finally, by how recipients agree and acquiesce to the care and obligation expressed both by donors and the goals of transplant medicine. The impacts of the mandate to live on the transformation of the subject are revealed in a more diffuse, less homogeneous fashion than are the impacts of the 'right to die'.

From Generation to Generation?

There is no question that clinical technique contributes to our moral sensibility (Brodwin 2000: 10). Yet, in his essay, 'From Generation to Generation', Thomas Laqueur (2000) emphasizes that the contemporary tensions surrounding the grounds of kinship connectedness, the meanings inherent in 'flesh of my flesh', did not arise recently with new (reproductive, or any other) biotechnologies. Rather, he notes, modern medical interventions simply provide one contemporary site for age-old family relations and obligations to be expressed. Biomedical technique in general, and live kidney donation in particular, provide the most powerful logic and one persuasive method to demonstrate love and care. Live kidney transfer extends the boundaries of what can be given and received. It expands the necessary explanatory work (both for recipients and donors) *to include* assumptions about the naturalness of life extension and the routine-ness of the generational direction of the gift and *to de-emphasize* the permanence of organ transfer and the finite character of this particular corporeal resource.

The biopolitics of relatedness emerged in the examples of our study subjects' words. Everyone who engages the world of transplant medicine makes moves to include and exclude, to name and rank those who will be considered worthy of giving and receiving (Carsten 2004: 180). Love, obligation, altruism, family solidarity, bodily risk, and assumptions about the naturalness of both mortality and transplantation play varying roles in individuals' moral reckonings.

The nature of obligation – from generation to generation – was perhaps expressed most clearly by a 54-year-old woman who donated a kidney to her boss and friend, who, she noted, was *like* family. Though she thought about the impact her donation would have on her own daughter and her daughter's children, she did not think about it for very long:

> He said that he had a kidney problem and that he was gonna look for a donor because he didn't want to be on dialysis. He gave me something to read, I think. But before then, I said yes … I've known X for a long, long time. We're like family – it wasn't a question. It was an easy decision. It prolongs the average recipient's life by about 16 years, I was told … My daughter has one kidney. She had cancer when she was six years old and they took a kidney out. That was 30 years ago. So, I know a little bit about it. Knowing that no harm was going to come to me and knowing that, if

down the line my daughter needed a kidney, she had some sort of back-up, I think that's what made it easy. I knew all of that ... I discussed it with my daughter and my son – what I was gonna do; how they felt about it. My daughter was totally for it and she and I talked about it and I didn't even have to say anything. She was the one who said, 'Well, I have two sons, you know, if I need a kidney ...' And that made me even more comfortable.

This donor's untroubled definitional move (Carsten 2004: 180) – 'we're like family' – named the significance of her bond to the patient and justified her gift as appropriate and unquestionable, even to her daughter who had only one kidney. This donor is not unique among those we spoke with in assuming that the transfer of her own kidney to another could instigate the 'natural' obligation of younger generations to donate to older kin and like-kin in the future. That always potential obligation of the body *of others* does not appear troublesome.

'Natural' limits to *life itself* have been obscured. The denial of a biologically ordained, ultimately inevitable death has become naturalized (Featherstone and Hepworth 1998: 156) both by clinical procedure and by an 'ethics of normalcy' (Rose 2001: 20) in which patients, families and providers all participate. The old question that Laqueur draws attention to – what are our obligations across generations? – has not disappeared. As ever, we must demonstrate the ways in which we care for the oldest members of our families and our societies. But, more than ever before, we are both being asked and demanding to share in those lives – in terms of bodily substance – through the medical procedures that are available.

Marilyn Strathern notes that medical interventions are also interventions 'into ideas' (1992: 5), and this is certainly true at the site of transplantation in later life. Just as the sonogram opened ways of seeing the foetus, its malformations and the idea of pre-birth intervention, as surrogacy opened up the idea of motherhood and family, and as cardiac surgery, the mechanical ventilator and emergency CPR changed ways of thinking about the risk of death, so, too, does the idea of organs moving from children to parents, spouses to each other or between friends or strangers open up the old issue of social and familial obligation to emerging biotechnical means of expression. We are concerned with the forms of being a biomedical subject, and the forms of caring, that are developing now. Once the idea of live kidney donation, especially from a child to a parent, has been conceived, expressed and made available to others (Strathern 1992: 33), it becomes normalized as part of the ethical landscape for the practice of obligation and for proactively contributing to a certain kind of somatic and ethical future – for the self, the family and subsequent generations. Live kidney transplant joins genetic, reproductive and pharmacological forms of social participation as one more technique linking ethics to intervention, embodiment to possibilities for selfhood and family responsibility, and the understanding of the arc of human life to clinical opportunity and consumption. Significant in this example is the medico-cultural scripting of transplant choice that elides to a high-stakes obligation in which the long-term impacts on generational relations cannot be foreseen.

Acknowledgements

This chapter is adapted from: Sharon R. Kaufman, Ann J. Russ and Janet K. Shim, 'Aged Bodies and Kinship Matters: The Ethical Field of Kidney Transplant', *American Ethnologist* 33(1): 81–99, 2006. We thank the American Anthropological Association and the University of California Press for permission to publish the article here in a shorter and slightly different form. The research on which this chapter is based was funded by the National Institutes of Health, National Institute on Aging, grant AG20962, to Sharon R. Kaufman, Principal Investigator. We thank Helen Lambert and Maryon McDonald for their invitation to present this work and for their help in refining the final version of this chapter. We are grateful to all the participants of the Cambridge Social Bodies Workshop for their enthusiastic response and thoughtful discussion.

Notes

1. According to the 2003 Annual Report of the US Organ Procurement and Transplantation Network, 1,684 persons in the US aged 65 and over received kidney transplants in 2003 (11.1% of the 15,135 total kidney transplants that year). The percentage of transplants for older persons has increased steadily since 1988, when 2.4% of transplants went to persons aged 65 and over. (http://www.optn.org/latest Data/rpt/data, accessed 29/12/2004: United Network for Organ Sharing, *Annual Report of the U.S. Organ Procurement and Transplantation Network and the Scientific Registry of Transplant Recipients* 2003). The situation is similar in Europe; in 1999 12.5% of all transplantations reported to the Eurotransplant registry were for persons over 65 (Schratzberger and Mayer 2003).
2. Although the actual number of older kidney recipients in the US (age 65 and over) has grown in the past two decades (from 213 in 1988 to 1,684 in 2003), the percentage of living donations to older persons has remained constant, at approximately 1% since 1989 (http://optn.org/latestData/rptData, accessed 29/12/2004). For all age groups combined, the sources of living donation have shifted in the past decade. For example, spouse donations have increased over time, from 4% in 1993 to 11% in 2002. Adult child donors

constituted 13% of donors in 1993 and 18% in 2002. Unrelated living donors have increased from 2.4% in 1993 to 17.6% in 2002, and are the group that has shown the greatest increase (http://www.optn.org/AR2003/209; http://www.optn.org/latest Data/rpt/data, accessed 29/12/2004: United Network for Organ Sharing, *Annual Report of the U.S. Organ Procurement and Transplantation Network and the Scientific Registry of Transplant Recipients: Transplant Data 1993–2002*, table 2.9).
3. The greatest increase in living donor type has been among unrelated donors (Mandal et al. 2003).
4. To preserve the anonymity of clinics, health professionals, patients and families, we do not name or describe our fieldsites, all of which are located in major medical centres in California.
5. For reports on donations to strangers, see Parker (2004) and Strom (2003). See also, Associated Press, 'Transplant Arranged via the Internet is Completed', 21 October, 2004 (http://www.nytimes.com/2004/10/21/national/21kidney, accessed 21/10/2004). For the worldwide traffic in illegal organs, see Rohter (2004). There is also a movie about the globalization of the illegal kidney market, *Dirty Pretty Things*.
6. Our use of 'ethical' here is influenced by the work of Foucault, especially his consideration of ethics as 'the considered practice of freedom' and the reflective part of freedom (Foucault 1997: 284–86).
7. The literature on subjectivity and intersubjectivity in the human sciences is rich and varied. Our thinking is grounded in the classic works of A.I. Hallowell, G.H. Mead, Maurice Merleau-Ponty and Alfred Schutz and is influenced by recent work in the human sciences on embodiment, phenomenology, self-making and social suffering and by ethnographic investigations of the social impacts of organ transplantation.
8. The ever-fraught gift relation is shaped with increasing regularity by poverty, coercion and the global market in 'bioavailable' kidney sellers and their needy and ready buyers (Cohen 2005). In the context of the illegal market the 'gift' is not the kidney, but is, for the seller, the temporary reprieve from debt or economic ruin or the promise of economic gain. Often that 'gift' is fraught with shame, chronic debility and severe ostracism for the coerced donor/seller (Scheper-Hughes 2004).
9. Most kidney transplant candidates (92.5%), regardless of age, are eligible for Medicare, which generally covers 80% of the cost of transplant surgery and 80% of the cost of anti-rejection medication for three years. In some cases the costs of anti-rejection medication are covered for as long as needed. In some cases Medicaid covers the costs of treatments for those ineligible for Medicare. And in some cases Medicaid covers the 20% of the costs of surgery and medications that Medicare does not cover. Medicaid and veterans' benefits vary considerably from state to state. Donor services are completely covered by Medicare. See: http://www.kidney.org/atoz; http://www.medicarerights.org and http://www.medicare.gov/publications. All accessed 2/12/2004. In our observations, fixed-income and low-income individuals often received transplants.

10. Our small, opportunistic interview sample (as of 31/12/2004) of 33 kidney recipients between the ages of 70 and 81 (24 men; 9 women) approximately mirrors the US national profile in terms of the ratio of living to cadaver organs transplanted: 19 to 15. (One person had two transplants from two live donors.) We also interviewed 28 persons (age 70 to 80) who were in the process of medical evaluation for a kidney transplant and spoke with eleven of their family members who had already offered to donate a kidney. Ethnically, our entire sample of recipients, donors, prospective recipients and their potential donors reflects the broad ethnic diversity found in metropolitan California: African American, Euro-American, Chinese American, Japanese American, Hispanic, Filipino, Samoan and immigrants from Afghanistan, China, Europe, the Philippines and Vietnam.

11. Medicare creates the possibility for transplant medicine in the US, and there is no doubt that federal reimbursement regulations structure and inhabit the culture of the clinic as well as individual patient treatments and outcomes. Yet a discourse about moral economy – about the relationship between differential access to transplant and Medicare – is rarely articulated. We thank Lawrence Cohen for a discussion of this topic. A specific analysis of the financing of transplants and its impacts on the ethical field is beyond the scope of this article and is the topic of future research.

12. Our opportunistic interview sample of donors included 17 women and 4 men, and it included the following relationships to recipients: 6 spouses (4 wives/2 husbands), 3 sisters, 8 daughters, 1 son, 1 nephew and 2 female friends. Within this small sample, wives offered and donated more often to husbands than husbands offered and donated to wives. Among siblings, women offered to donate to their parents before their brothers did and they did so more easily, enthusiastically and insistently. They were more adamant than their brothers that a parent accept their gift and they expressed no reservations about giving. Their brothers, we were told, did not offer as quickly and were more reticent about giving at all. Our sample to date is too small to comment on the role gender plays in living donation. United Network Organ Sharing statistics show that men receive about one-third more kidneys (both cadaveric and living) than do women because men have more kidney disease. (In 2004, 9,574 men received kidney transplants; 6,428 women received kidneys. See: http://www.optn.org/latestData/rptData.asp, accessed 8/7/2005. Greater kidney disease in men also prevents them from becoming donors. This biomedical fact needs to be considered in an analysis of gender and the tyranny of the gift.

References

Arney, William Ray and Bernard J. Bergen. 1984. *Medicine and the Management of Living*. Chicago: University of Chicago Press.

Biehl, Joao. 2004. 'Life of the Mind: The Interface of Psychopharmaceuticals, Domestic Economies, and Social Abandonment', *American Ethnologist* 31(4): 903–10.

Biehl, Joao, Denise Coutinho and Ana Luzia Outeiro. 2001. 'Technology and Affect: HIV/AIDS Testing in Brazil', *Culture, Medicine and Psychiatry* 25: 87–129.

Brodwin, Paul. 2000. 'Introduction', in Paul E. Brodwin (ed.), *Biotechnology and Culture*. Bloomington: Indiana University Press.

Carsten, Janet. 2001. 'Substantivism, Antisubstantivism, and Anti-antisubstantivism', in Sarah Franklin and Susan McKinnon (eds), *Relative Values: Reconfiguring Kinship Studies*. Durham: Duke University Press.

———— 2004. *After Kinship*. Cambridge: Cambridge University Press.

Chkhotua, Archil B., Tirza Klein, Eti Shabtai, Alexander Yussim, Nathan Bar-Nathan, Ezra Shaharabani, Shmariahu Lustig and Eytan Mor. 2003. 'Kidney Transplantation from Living-Unrelated Donors: Comparison of Outcome with Living-Related and Cadaveric Transplants Under Current Immunosuppresive Protocols', *Urology* 62: 1002–06.

Clarke, Adele, Janet Shim, Laura Mamo, Jennifer Fosket and Jennifer Fishman. 2003. 'Biomedicalization: Technoscientific Transformations of Health, Illness, and U.S. Biomedicine', *American Sociological Review* 68: 161–94.

Cohen, Lawrence. 2005. 'Operability, Bioavailability, and Exception', in Aihwa Ong and Stephen J. Collier (eds), *Global Assemblages*. Malden, MA: Blackwell.

Csordas, Thomas. 1990. 'Embodiment as a Paradigm for Anthropology'. *Ethos* 18: 5–47.

———— 1994. 'Introduction: The Body as Representation and Being-in-the-World', in Thomas Csordas (ed.), *Embodiment and Experience: The Existential Ground of Culture and Self*. Cambridge: Cambridge University Press.

Davis, Connie L. 2004. 'Evaluation of the Living Kidney Donor: Current Perspectives', *American Journal of Kidney Disease* 43(3): 508–30.

Davis, Connie L. and Francis L. Delmonico. 2005. 'Living-Donor Kidney Transplantation: A Review of the Current Practices for the Live Donor', *Journal of the American Society of Nephrology* 16(7): 2098–110.

Dowd, Maureen. 2003. 'Our Own Warrior Princess', *New York Times* (1 June).

Edwards, Jeanette and Marilyn Strathern. 2000. 'Including Our Own', in Janet Carsten (ed.), *Cultures of Relatedness: New Approaches to the Study of Kinship*. Cambridge: Cambridge University Press.

Estes, Carroll L. and Elizabeth A. Binney. 1989. 'The Biomedicalization of Aging: Dangers and Dilemmas', *The Gerontologist* 29(5): 587–97.

Featherstone, Mike and Mike Hepworth. 1998. 'Ageing, the Lifecourse, and the Sociology of Embodiment', in Graham Scambler and Paul Higgs (eds), *Modernity, Medicine and Health*. London: Routledge.

Foucault, Michel. 1980. *The History of Sexuality, Volume I, An Introduction*. New York: Vintage Books.

———— 1983. 'The Subject and Power', in Hubert L. Dreyfus and Paul Rabinow (eds), *Michel Foucault: Beyond Structuralism and Hermeneutics*, 2nd Edition Chicago: University of Chicago Press.

———— 1997. *Ethics, Subjectivity and Truth*, ed. Paul Rabinow. New York: The New Press.

Fox, Renee C. 1996. 'Afterthoughts: Continuing Reflections on Organ Transplantation', in, Stuart J. Youngner, Renee C. Fox and Lawrence J.

O'Connell (eds), *Organ Transplantation: Meanings and Realities*. Madison: University of Wisconsin Press.

Fox, Renee C. and Judith P. Swazey. 1992. *Spare Parts: Organ Replacement in American Society*. New York: Oxford University Press.

——— 2002. *The Courage to Fail*. New Brunswick: Transaction Publishers.

Franklin, Sarah. 1997. *Embodied Progress: A Cultural Account of Assisted Conception*. London: Routledge.

——— 2000. 'Life Itself: Global Nature and the Genetic Imaginary', in Sarah Franklin, Celia Lury, and Jackie Stacey (eds), *Global Nature, Global Culture*. Thousand Oaks, CA: Sage.

——— 2001. 'Biologization Revisited: Kinship Theory in the Context of New Biologies', in Sarah Franklin and Susan McKinnon (eds), *Relative Values: Reconfiguring Kinship Studies*. Durham: Duke University Press.

Franklin, Sarah and Susan McKinnon. 2001. 'Introduction', in Sarah Franklin and Susan McKinnon (eds), *Relative Values: Reconfiguring Kinship Studies*. Durham: Duke University Press.

Grady, Denise. 2001. 'Transplant Frontiers: Healthy Give Organs to Dying, Raising Issue of Risk and Ethics', *New York Times* (24 June): 1, 16.

Katz, Stephen and Barbara Marshall. 2003. 'New Sex for Old: Lifestyle, Consumerism, and the Ethics of Aging Well', *Journal of Aging Studies* 17: 3–16.

Kaufman, Sharon R. 1994. 'Old Age, Disease, and the Discourse on Risk', *Medical Anthropology Quarterly* 8(4): 430–47.

——— 2005. *And a Time to Die: How American Hospitals Shape the End of Life*. New York: Scribner.

Kaufman, Sharon R., Janet K. Shim and Ann J. Russ. 2004. 'Revisiting the Biomedicalization of Aging: Clinical Trends and Ethical Challenges', *The Gerontologist* 44(6): 731–38.

Kleinman, Arthur. 1997. '"Everything That Really Matters": Social Suffering, Subjectivity, and the Remaking of Human Experience in a Disordering World', *Harvard Theological Review* 90(3): 315–35.

Koenig, Barbara. 1988. 'The Technological Imperative in Medical Practice: The Social Creation of a "Routine" Practice', in Margaret Lock and Deborah Gordon (eds), *Biomedicine Examined*. Boston: Kluwer.

Laqueur, Thomas W. 2000. 'From Generation to Generation', in Paul E. Brodwin (ed.), *Biotechnology and Culture*. Bloomington: Indiana University Press.

Lock, Margaret. 1996. 'Deadly Disputes: Ideologies and Brain Death in Japan', in Stuart J. Youngner, Renee C. Fox and Lawrence J. O'Connell (eds), *Organ Transplantation: Meanings and Realities*. Madison: University of Wisconsin Press.

——— 2002. *Twice Dead: Organ Transplants and the Reinvention of Death*. Berkeley: University of California Press.

Mandal, Aloke K., Jon J. Snyder, David T. Gilbertson, Allan J. Collins and John R. Silkensen. 2003. 'Does Cadaveric Donor Renal Transplantation Ever Provide Better Outcomes than Live-Donor Renal Transplantation?', *Transplantation* 75(4): 494–500.

Mauss, Marcel. 1967. *The Gift: Forms and Functions of Exchange in Archaic Societies*, translated by Ian Cunnison. New York: Norton.

Merleau-Ponty, Maurice. 1962. *Phenomenology of Perception*, translated by Colin Smith. London: Routledge and Kegan Paul.

Nolan, Marie T., Benita Walton-Moss, Laura Taylor and Kathryn Dane. 2004. 'Living kidney Donor Decision-Making: State of the Science and Directions for Future Research', *Progress in Transplantation* 14(3): 201–9.

Parker, Ian. 2004. 'The Gift'. *The New Yorker* (2 August): 54–63.

Parsons, Talcott. 1951. *The Social System*. Glencoe, IL: Free Press.

——— 1975. 'The Sick Role and the Role of the Physician Reconsidered', *Millbank Memorial Fund Quarterly* 53 (Summer): 257–78.

President's Council on Bioethics. 2003. 'Beyond Therapy: Biotechnology and the Pursuit of Happiness'. Electronic document, http://www.bioethics.gov/reports/beyondtherapy

Rabinow, Paul. 1996. *Essays on the Anthropology of Reason*. Princeton, NJ: Princeton University Press.

——— 2000. 'Epochs, Presents, Events', in Margaret Lock, Allan Young and Alberto Cambrosio (eds), *Living and Working with the New Medical Technologies*. Cambridge: Cambridge University Press.

Rabinow, Paul and Nikolas Rose. 2003. 'Thoughts on the Concept of Biopower Today'. Conference paper, *Vital Politics*, London School of Economics and Political Science (5–7 September), London.

Rapp, Rayna. 1999. *Testing Women, Testing the Fetus: The Social Impact of Amniocentesis in America*. New York: Routledge.

Rohter, Larry. 2004. 'Tracking the Sale of a Kidney on a Path of Poverty and Hope', *New York Times* (23 May): A1.

Rose, Nikolas. 2001. 'The Politics of Life Itself'. *Theory, Culture and Society* 18(6): 1–30.

Rose, Nikolas and Carlos Novas. 2005. 'Biological Citizenship', in Aihwa Ong and Stephen J. Collier (eds), *Global Assemblages*. Malden, MA: Blackwell.

Scheper-Hughes, Nancy. 2004. 'The Last Commodity: Post-Human Ethics and the Global Traffic in "Fresh" Organs', in Aihwa Ong and Stephen J. Collier (eds), *Global Assemblages*. Malden, MA: Blackwell.

Schneider, David. 1980. *American Kinship: A Cultural Account*, 2nd edition. Chicago: University of Chicago Press.

Schratzberger, Gabriele and Gert Mayer. 2003. 'Age and Renal Transplantation: An Interim Analysis', *Nephrology Dialysis Transplantation* 18: 471–76.

Segev, Dorry L., Sommer E. Gentry, Daniel S. Warren, Brigitte Reeb and Robert A. Montgomery. 2005. 'Kidney Paired Donation and Optimizing the Use of Live Donor Organs', *Journal of the American Medical Association* 293: 1883–90.

Sharp, Lesley A. 1995. 'Organ Transplantation as a Transformative Experience: Anthropological Insights into the Restructuring of the Self', *Medical Anthropology Quarterly* 9(3): 357–89.

——— 2001. 'Commodified Kin: Death, Mourning, and Competing Claims on the Bodies of Organ Donors in the United States', *American Anthropologist* 103(1): 112–33.

Siminoff, Laura A. and Kata Chillag. 1999. 'The Fallacy of the "Gift of Life"', *Hastings Center Report* 29(6): 34–41.

Strathern, Marilyn. 1992. *Reproducing the Future: Anthropology, Kinship, and the New Reproductive Technologies*. New York: Routledge.

Strom, Stephanie. 2003. 'An Organ Donor's Generosity Raises the Question of How Much is Too Much', *New York Times* (17 August): A12.

Thompson, Charis. 2001. 'Strategic Naturalizing: Kinship in an Infertility Clinic', in Sarah Franklin and Susan McKinnon (eds), *Relative Values: Reconfiguring Kinship Studies*. Durham: Duke University Press.

United Network for Organ Sharing. 2003. *Annual Report of the U.S. Organ Procurement and Transplantation Network and the Scientific Registry of Transplant Recipients: Transplant Data 1993–2002*. Rockville, MD.

Wolfe, Robert A., Valarie B. Ashby, Edgar L. Milford, Akinlolu O. Ojo, Robert E. Ettenger, Lawrence Y.C. Agodoa, Philip J. Held and Friedrich K. Port. 1999. 'Comparison of Mortality in All Patients on Dialysis, Patients on Dialysis Awaiting Transplantation, and Recipients of a First Cadaveric Transplant', *New England Journal of Medicine* 341: 1725–30.

Chapter 2

ANATOMIZING CONFLICT –
ACCOMMODATING HUMAN REMAINS

Maja Petrović-Šteger

Anthropological accounts of the body and person have recently begun to undergo a reorientation, with a shift of attention from the body understood as an integral unit to an idea of the body as sometimes made up by dead or dismembered body parts. While this study joins the debate on body parts (see Lock 2002; Scheper-Hughes and Wacquant 2003; also Kaufman et al., in this volume), it does not, however, address the phenomenon of the booming market in human organs destined for transplantation, whether to extend lives or modify bodies. Instead, the bodies and body parts emerging at the centre of this analysis are medically unusable, human remains whose use-value is clearly extinguished, but which, despite their 'uselessness', continue to offer sites for new scientific, medical and technical interventions.

More specifically, the chapter will examine a range of material practices and rhetorical strategies constructed around bones and other human remains in postconflict Serbia. In present-day Serbia (as in other postconflict areas of former Yugoslavia), the recovery, identification and return of shattered bodies and body parts (the common fate of bodies in war) to families has been identified by divergent political interests as potentially healing or restorative. Remains are articulated in such a way as to build and sustain communities in grief and to posit continuities between past and present. Yet the forms in which body parts circulate in these situations – as a means for reconciliation, as commodities, as private mementos, and as DNA-coded information – are, as this account reveals, more various than the official narratives of attribution and assignment suggest.

Drawing on fieldwork carried out in 2003/04 in Serbia and Bosnia-Hezegovina.[1] the text sets out to explore how narratives of the conflict, enacted through human remains and their evidentiary traces, play themselves out in postconflict practices of intervention into, and collection and classification of, body parts. My analysis shows how war, as a highly sensitive period in a group's cultural memory, becomes medicalized, lending itself to appropriation, in peacetime, through the operation of a large-scale forensic and generally scientific apparatus. In this process, the study suggests, arguments over the meaning of human remains can serve as metonyms for debates over the justification for conflict and the subsequent negotiation of the postconflict political order.

Human Remains in Serbia

The political deployment of mourning has a history in the former Yugoslav republics going back well beyond its most immediate occasion in the conflicts of the 1990s. Typically, the rhetoric of grief involves the invocation of a symbolic form, or a symbolization of the materiality, of the dead body, which is used to rally identification with a historical heritage or putative ethnic identity. In the late 1980s, political propaganda in Serbia (as well as in Croatia, Bosnia and Herzegovina (BiH), and Macedonia to a lesser degree) repeatedly appealed to certain legitimating myths of descent in seeking to make up a national *ethnos* or serviceable political community. Such a community was held together rhetorically by the postulation of biologically traceable kinship relations, as imaged in dead bodies, graveyards and bones. Proponents of war in Serbia invoked the ancient motif of bones – magical caskets, according to folk religion and local mythology, lodging the soul of Serbian people – in order to activate pro-war political feelings and to make territorial claims.

The importance of customs, ideas and practices about death, burial and the proper processes of bereavement in Serbia has been amply documented since the early twentieth century.[2] Anthropological and ethnological testimonies show, for example, that it used to be a common Orthodox practice to exhume the dead a certain time after burial (three, five, or seven years), to then wash the bones [3] and to rebury them with a special liturgy.[4] Thus, two years before the commemoration of the 600th anniversary of the 1389 Battle of Kosovo,[5] in which Serbia lost its autonomy to the Ottomans with the death of Prince Lazar, Lazar's remains were taken on a ceremonial tour[6] of Serbia, with the majority of the locals extending a proud welcome to the procession. The remains were first taken from the patriarchate in Belgrade (the latest of Lazar's various abodes over the last six centuries) and then borne to rest temporarily in monasteries in all the regions where the Serbs had churches.[7] Critics of this act claim that it was the highly political portage of the Prince's bones that drew up the boundaries of 'Great Serbia', on the principle that 'wherever the bones of Serbian ancestors lie, that is Serbian land'. Though evidently anachronistic, the archaic principle served crucially to forge pro-war political alliances in the Yugoslav wars of the 1990s. In other words, certain territories figured within imaginary conceptions of the land as being specifically imbued with mythic value, and unnameable ancestors interred in those places started to function as real persons as they retrieved or symbolically reconquered the ground in which they lay.

Importantly, the rituals and practices as documented did not exclusively bear out the doctrine of the Orthodox Church, but formed part of Serbian popular religion[8] or the so-called 'folk religion of Serbs'.[9] Unwritten rules – about how to treat the deceased – have been passed from one generation to another through oral folkways as a form of local knowledge about death.

Spending a lot of time on the grave-sites in 2003/04, I had the chance regularly to observe how 'folk religious practices' inform contemporary public memorial techniques. A comparable etiquette of administering burial properly seemed shared by religious and supposedly secular methods. Family members in Serbia buried their dead with regular elaborate prayers and mortuary feasts held according to a fixed temporal pattern (on the first-falling Saturday or seven days after death, after forty days, within six months and one year); these ceremonies were understood to propitiate the dead and ensure their contentment, acting, further, as a failsafe against their becoming a vampire (see Jovanović 2002).[10] Many of my respondents assured me that they still held tightly to received patterns. Serbian funerary practices, as both documented and extant, are informed by the belief that the souls of the deceased remain alive so long as any part of his/her body, especially their bones, persist on the earth, given that the skeleton is the abode of the soul (Čajkanović 1994: 417). Whether or not contemporary Serbians would actually profess these beliefs, they remain alive in turn in people's remembrances of their forebears, in their imprecations and in their solemn recitation of religious incantations.

Needless to say, the period of my fieldwork was one during which I both observed as an ethnographer and was observed. I would regularly see a woman, in her fifties and always smartly dressed, resting under a slender white birch overshadowing a family vault in one of Belgrade's city cemeteries. One day this woman introduced herself as Ružica and asked me whom was I visiting. It turned out that she had spotted me as I habitually walked around, rather than keeping myself to only one grave. When I told her about the theme of my research, she let on that her whole family was buried in her dark marble family vault: both her parents, her father's parents, and her grandfather's two sisters. She had recently buried also her 26-year-old son there, she confided, indicating the darker heaps of earth around the vault and its newly planted flowers. 'My son died whilst serving his military conscription', she said somewhat stoically, seeming disinclined to offer further information. As we stood in silence for a while, I noticed that different surnames had been engraved into the ledger of the family vault. The names of the dead were (or could have been) of both Croat and Serbian origin. Ružica explained that in the early 1990s she had got a phone call informing her that everyone in her son's squadron had perished in a military operation, most likely he included. The army, though, had never officially confirmed her son's death. After a fruitless wait and period of searching for her son, or his remains, lasting eleven years, she had engraved his name on the ledger – thus symbolically burying him – and planted some new flowers, last month, 'so that at least his soul could find rest'.[11] In a lugubrious, but also slightly hectoring way, she intoned, 'What counts is showing respect'. Properly administered burial still counted – even if it meant people conducting obsequies over empty tombs.

But to return to the early 1990s, it was not so much the joining with ancestral mourning practices, or traditional Serbian (re)burials,[12] as the rediscovery of Serbia's World War II dead that was crucial in igniting warfare in 1991. During the latter part of the Second World War, in which Yugoslavia resisted German occupation, war atrocities were inflicted on Yugoslavs not just by the Reich but by groups of their fellow countrymen. More than a million Yugoslavs died in the war, mostly at their compatriots' hands. Committed by numerous groups, and on all sides (the perpetrators being Croatian fascists known as Ustaša, against Serbs; partisans, against fascists; Serbian royalists – Četniks – against partisans, Muslims and other Croats, etc.), these multiple massacres, mention of which was silenced during Tito's regime,[13] became the object of revisionist histories, usually political and nationalistic, in the late 1980s. After Tito's death,[14] the once-socialist regime of former Yugoslavia came under increasing strain, with additional care for corpses and graves being seen as essential for religious and ethnic renewal. A number of exhumations and reburials were televised, and viewed by large audiences across the whole of the former country.

Such images – of people digging out bones from caves, bagging them and turning these plastic sheaths to the light (cf. Verdery 1999) – thus became the sources of mutual recrimination and demonization among ethnic groups of Yugoslavs (cf. Denich 1994; Hayden 1994; Ballinger 2003). Both in folk rhetoric and in political speeches, bones and body parts of Second World War victims that had been thrown into caves, buried in shallow graves, or simply left to rot, served to mark and spatialize the new state borders of the 1990s, further differentiating 'true' or 'national' kinship[15] and ethnic ties from more general modes of relatedness based on citizenship. The reburial of exhumed remains played an essential role in suturing cultural nationalist understandings of the dead body to religious mortuary forms.

Along with documentary films on Second World War genocides, a number of highly regarded writers published in the early 1990s works on Yugoslavia's Second World War history, with the different interpretations, or manipulation of such, of each arousing dormant emotions which again got caught up in strategies of attributing guilt and accountability. According to Verdery (1999), the projection of unmarked corpses, or 'nameless bodies', was particularly effective in exciting these animosities. Nationalist demagogues used the ethnic crimes of the past to fuel a new cycle of ethnic violence. In a bid to attract honour to the idea of fighting for the Serbian nation in the 1990s, war apologists described a citizen as someone with a proper degree of respect for his recent and distant ancestors. As well as being mediated through death, in this process, civic culture was invested with distinctively masculine values: individuals who fought for 'just cause' were given terrestrial and national rights in return for their efforts on the battlefield. Acts performed in the name of an ancestral principle found justification in an idiom of anteriority and claims over land and property. In Judith Butler's terms, because actions could be articulated as drawing on a symbolic dimension, they were able to lay claim to a universal force (see Butler 2000: 44). Isolated actions or skirmishes in the war, including war crimes, often clothed themselves in the name of a principle that simultaneously elevated and sanctioned them. Those Serbs that were pro-war, moreover, expressed their fidelity to an ideology of historical Serbia, and their proximity to political power, through spatial markers and land ownership claims.

The dead body was key in mobilizing this nexus of relations between power, land title and historical entitlement; but alongside the corpse's effectivity in a public symbolic register during the wars of the 1990s, we should also note the alleged flourishing of a particular material practice around corpses at this time. This was the bartering of body parts and whole corpses between Serbs, Croats, Bosnians and later Kosovars for what my respondents called 'emotional' and 'spiritual' reasons. Although no official documents discuss this trade, a number of my interviewees (Serbs, Bosniaks and Croats) admitted that they were either aware of such trafficking, or had even taken part in it. This return of the dead body also played up to a potent and readily politicized set of popular expectations. In bringing the body parts of dead relatives back to the soil where they 'belonged', those involved in the trade understood themselves to be restoring their own spirits, and those of their ancestors, to peace.

Human Remains in Times of Postconflict

This study is not concerned, however, simply with the atavistic dimension of people's desire for the repatriation of their dead relatives, as this is sought on an informal level beneath the oversight of official or state-sector processes of administration. On the contrary, in the negotiation of the postconflict Serbian order, what might be thought an atavistic, 'non- or pre-modern' practice finds a counterpart in the use of these same remains to reinflect or reconstruct ties and ideas of kinship within (and between) statelevel entities. Where once ancestors were iconically invoked as figures conjuring national kinship, now a symbolic body politic is taking shape around the bodies of those who died in the past decade. This is more than a matter of membership of the 1990s war dead, say, among one's immediate family bestowing legitimacy on a citizen or national subject. In addition to these popular forms of recognition, a number of public sectors, NGO and civil society organizations are active in resuscitating an idea of national participation through their solicitation of families' involvement in scientific and bureaucratic processes of bodily return. In the view of the international community, the peoples of the former Yugoslavia commit themselves to a course of healing by beginning to reckon with the evidence of war crimes. A Serbian public has begun to reconcile itself to the postwar polity by consenting to take part in internationally run programmes of repatriation; while politicians and others have sought to frame the terms of Serbia's guilt, or its relation to Europe and the world, by recourse to the work of agencies dealing with the dead. The ICMP (International Commission on Missing Persons), created in 1996 in Lyon by the G7 in order to address the issue of persons missing as a result of the conflicts of the 1990s, plays an important role in these surrogate processes.

Following the cessation of hostilities and the signing of the Dayton Peace Accord, the ICMP[16] started its work first in Bosnia and Herzegovina, before extending its activities to other postconflict regions. Estimating that more than 40,000 persons remained unaccounted for across the former Yugoslavia, the Commission established, under a 'strictly humanitarian mandate', a 'mission to bring relief' to the families of the missing, regardless of religious, national or ethnic origin. The organization's remit in retrieving remains on families' behalf connects with the wider societal effort to frame the legal terms of accountability and reconciliation, as transacted in bodies, rights and across a range of other political and civil contexts.[17]

The Commission uses DNA-led methods, developed by its own forensic specialists, to identify the remains it exhumes. In order for DNA to figure in the identification of exhumed bodies, however, DNA profiles must be taken in the form of blood samples from the family members of a missing person, and compared to the profile abstracted from recovered body parts and bones. This sampling of a population necessitated a major bureaucratic or 'public health' campaign spanning, and crucially 'reconnecting', the whole of the former Yugoslavia. Compiling a dossier, or rather a satisfactorily comprehensive database, of samples thus required the cooperation of a number of previously hostile groups, including the grieving wives, mothers and daughters of the deceased, other self-declared victims, possible perpetrators and governmental bodies; such groups had to exchange information for war crimes localities to be determined and, potentially, for families to accept remains and achieve closure. Should locals have held back from reporting their missing,[18] or, further, from providing information as to the possible whereabouts of mass graves (and identity of perpetrators), foreign experts would not know where to search, frustrating identification efforts from the outset.

It took some time, then, for the ICMP and related organizations[19] to generate the primary source of their data – that is, the register and inventory of missing persons – that would form the basis of their future work. Generally speaking, the local people I worked with, both Serbs and Bosniaks, tended to be very ready to offer accounts of how they were encouraged to 'face their experience of the wars by voicing their opinions' -what they sometimes called their 'pressure situations'. But if people have largely dispensed with any embarrassment in narrating their suffering as caused by the war, it remains harder to summon up the courage to consider its roots. These roots, understandably, are held to be complex and multiple; and although many have sought to pin down the major factors precipitating the conflicts, most local writing and commentary is regarded by the international audience as nationalistic, pro-war or one-sided etc. Likewise, local Serbs often read the international intervention in the 1990s in Yugoslavia as 'aggressive, usurpatory, belated and economically-driven'. As the initial operations of the ICMP consisted primarily in gathering information to be used as legal evidence in The Hague court, they were welcomed in Bosnia by representatives of different ethnicities but tended to be repudiated by many in Serbia (either through mockery or neglect). Undergirded by such a social dynamic, the objectives and work of the ICMP were initially met with great suspicion fostered by the obvious international character of the Commission as an intragovernmental organization founded by Bill Clinton in 1996. The initial relationships between the Commission and the local population, in the words of an ICMP officer, were: 'Frustrating. People thought we were an aid agency that would provide them with financial help. But we didn't come here to distribute money. We didn't have much money anyway. We said that we had come to find their war missing'.

Another fact contributed to the tension between the locals (Serbs, Bosniaks and Croats) and the ICMP. At the time just after the signing of the Dayton Accord, despite the fact that positive collaboration between locals and foreign volunteers and experts represented a prerequisite for the ICMP's work, it was believed that locals should be excluded from the professional processes of remains' retrieval. As the mass graves found were rarely primary, but usually secondary or tertiary sites, the 'ethnic' origin of their victims was not clear, with the possibility existing that graves held people from all three sides of the conflict. The ICMP wanted therefore to ensure that no specific local nationality would be represented in (or could be seen as favoured by or allied with) the ICMP working team. Yet this decision provoked dissatisfaction among the people personally involved in searching for their loved ones' remains. I recorded many stories of frustration, in which locals claimed that they had initially felt 'cheated, as no one had let them come close to the excavation sites', prohibiting them, moreover, from dealing with the remains as they felt proper. A great number of the ICMP workers whom I interviewed stated that they were advised not to mingle much with the locals, as their job was to collect data that very well might be destined for use in criminal justice courts. Interestingly, many found my questions as to their knowledge of the region, or the nature of their contact with locals, misplaced or emotionally redundant. A middle-aged archaeologist once commented somewhat tiredly: 'I have been here for more than 5 years. No, I don't speak the local language, but should I have to? I came here to find people's remains, and not to amuse them with my language capabilities or the lack of such'.

Another forensic scientist followed up by commenting on the nature of the ICMP:

> Yes, but that's what is truly strange here. The policy of the ICMP is to have a fractured process – different experts work on different stages of identification. And this separates us from families. Where I worked before, in El Salvador, Zimbabwe, Argentina … we were involved in the whole process … from visiting the families and asking them for consent to dig up the remains, to handing the remains back to families. That meant that we had to communicate with them. That is much better, I think. Being involved in the whole process allows you at least to see the outcome – the return of the body to the family. With that you can somehow close the process for yourself. But here … no. You are always only digging and digging and digging … and you rarely find out where the bodies you worked on end up. But then … I have never worked at a site such as this before. Here we have thousands and thousands of bodies to excavate. Even if you wanted to, it

would be impossible to supervise the whole process from beginning to the end, because it is just too complicated and too big.

Another ICMP expert, a tiny woman in her forties, added: 'It's true. We might not communicate with them, but we help them nevertheless. We detect and identify those whom they miss. We help them reach justice'.

With time, the ICMP started helping the families of missing, also offering financial support, training and technical assistance in writing reparation claims, and serving as a forum for the organization of meetings, conferences and all sorts of special events. Selma, an ICMP regional coordinator born in Bosnia-Herzegovina, who struck me as a brisk and very resolute woman, explained:

> We think it's crucial to encourage people to play a real role in resolving issues around their missing. To report that somebody is missing already counts for a lot, but it is not enough. People need to be more active. They have to speak so that we can hear them. They have to talk about their cases and build up public awareness. And they should do that in an organised way. Nobody can possibly feel better when mourning alone. We can help them feel better … We can't always find their missing loved ones, but we are trying to help them by setting [relatives] up in networks and organising roundtables in which they can participate … if they want. Many people have told me that these occasions [conferences, workshops] really help, and make them feel decent again … help them bear their loss. …Yes, I do think we are helping them. You saw it yourself. It's very often the case that after a woman [it is usually women who assert victims' identification] identifies her missing person she stays with us as an organisation. Many women have become seriously involved in the ICMP workshops and organisations. … I guess, that once they cease to be dependent on us [regarding the information about deceased or the missing] they are more able to get a hold on their emotions, and they usually want then to throw themselves into the other activities … But of course, it is our duty to mobilise them in the first case. And the best way to do that is to invite them for a DNA testing. It is terribly hard to come into contact with them otherwise. You know, we're in touch with a lot of people already who are fine with coming forward, but we also have to go out and look for those who are sitting on their own at home seek and don't know who to ask for help. The fact that these people can meet our workers, and see that they really are trustworthy, and that they are locals themselves, whilst donating blood, is really important. It seems it really helps. And people have learned with time *how miraculous the DNA techniques can be* anyway. (my emphasis)

As over the years the ICMP's policies have changed, so have local people's evaluation of their work. Locals, such as Selma, have now been invited to join the ICMP in a professional capacity, while the humanitarian aspect of the organization has been promoted with increasing salience:

> The reason why the ICMP was founded was to put a stop to the trafficking of bodies and to facilitate the exchange of information about bodies and grave sites for financial and diplomatic reasons. We wanted to set the terms under which this exchange could become a humanitarian act. The point was to encourage the Serbs to help the Bosniaks by admitting what they knew about their missing persons, and for the Bosniaks to help the Serbs, and the Serbs to help Kosovars, or Croats help Serbs, and so forth. We knew that the circle of information always existed, and we did not want to break it. We wanted just to place it on a different, let's say, moral basis. (ICMP officer from the Belgrade office)

And so the retrieval and identification of remains has lent itself to inter-ethnic reconciliation as much as it has to more personal healing; the process, in both materially and metaphorically reassembling previously dishonoured and ravaged remains, has allowed families to perform funerals and commemorate deaths – reinscribing their losses into a comprehensible sequence.

Mint Fields and Mass Graveyards

Between 2003 and 2004 I attended a number of exhumations and excavations of human remains in both Serbia and Bosnia-Herzegovina (BiH). The bodies found within various mass graves had suffered a range of disfiguring injuries. When I recall one of these sites, the first image that comes into my mind is of a swaying field of wild mint. Consumed with nerves and genuinely ignorant of what I was going to see, the first time I visited a grave-site I was overwhelmed against all expectations by the fresh smell of mint carried on a July wind. It did not feel like walking across a mass grave to push through the soft, unmown grass heaving with buttercups. Only past the military tent, the bulldozer and other heavy equipment, did the mint give way to a more pervasive odour of decay. The yellow ribbon under which I ducked to enter the site read: 'Police – Crime scene – Entrance prohibited'. Approximately 16 metres in length, 3 metres wide and 2.5 metres deep, with a ramped-up base identified as the grave's western end, the grave held the remains of fifty two individuals. Viewed from the side, the plot had the shape of a wedge. The presence of fragments from earlier deposits, and the remains' stratigraphic context, suggested that the grave was a secondary one. As I came onto the site, archaeologists were removing soil from squares marked out in a grid, sieving it for skeletal and other debris. Skulls were often smashed and lay at odd angles. A senior forensic archaeologist, a silver-headed man who, it seemed to me, sat in the grave as comfortably as he might in a sandpit, showed me one of the skulls, where the wound to the right frontal bone, or forehead, bore small, circular edges – the signature-mark of a projectile. Commingled burned and fragmented remains had been separated with great care from whole corpses. The soft, knobby and grumose tissues of the body were kept to one side of the skeleton by the dead person's clothes. Following the standard postmortem protocol, clothing was removed, handwashed, described and photographed. Notwithstanding the close observance of this procedure, adipocere, a brownish-white, soap-like material, was everywhere.

It took me a while to strike up conversation with an older, heavily moustached local worker, who seemed to have been digging up and sifting layers of soil for an eternity. A cacophony of birdsong and drone of heavy diggers made it difficult for us to hear one another. I had to step down practically into the grave if I wanted to speak to him. The man, hands caked with a heavy layer of dirt, smiled at me and said in a tired way: 'You'd get used to it. Everyone gets used to it. It's hard at first, but then you forget about it quite soon, and think of something else while you dig. When we have breaks we even joke about this stuff. It's only life'. I nodded and asked in return what they joked about. Did he talk about his work on the grave-sites with his family? He looked at me seriously and said: 'They don't know what I do. Of course they don't. They know that I work for the ICMP, and that I bring good money back home, but I'd certainly not go around telling that I was the person who digs up and cleans the bones here'. The man then turned his back on me as if offended and embarrassed at the same time, and went on digging. I stepped out of his cave, and into another, that of a forensic archaeologist in her late forties, who was smoking a cigarette while dusting off and counting the hand bones of a skeleton facing skull-down on the ground. Another archaeologist was carefully removing decaying tissue from a newly-discovered body with a scalpel. He commented:

> This one is in quite a rough state. Generally speaking, when a body is buried in a shallow grave, it's subject not only to the attentions of many insects and animals, but also affected by seasonal fluctuations in temperature. Hmm, the surface of the articular region of the hipbone, is pitted heavily … and … you see … it has uneven borders. He must have been over sixty years old.

Finding the smell, sights and information all too much to take in at once, I made a determined effort at least to remember what I saw. I consciously tried to retain archaeologists' comments, the shape of collarbones, and to memorize dental record files, the colours of clothing, images of bits of broken mirror, and of the few coins that were sparkling in the pit. An archaeologist showed me a cache of items he had found – identification documents encased in plastic stapled to the collar of one man's clothes, and a set of prayer beads (*tespih*),[20] together with a SIM mobile phone card. On retrieval, each set of remains was placed in a fresh, white body bag, labelled and carried to the storage tent to await autopsy. The impressions formed by the teeth of a digging and loading shovel sat on a large wheeled machine testified to the effort that had gone into constructing the grave. The whole mint field was anyway an irregularly shaped area, the surviving mark of disturbance associated with the stockpiling of removed soil, or so I was told.

The images of those things I could not grasp immediately but hoped to remember – prostrate skeletons, putrefied tissues, a powerful whiff of ammonia, clothes,[21] tyre marks – all appeared to be perfectly intelligible, however, to the rest of the crew working on the site. What I apprehended as sensory impressions, others treated as evidence, proofs fitting into an interpretive framework designed to yield an analytical synopsis of the exhumation in a medico-legal context. The whole rationale of the examination and codification of the fragments and body parts lay in the legal meanings they would acquire as medical evidence. But I could not help feeling that, as well as helping to detect and prosecute crimes, forensic experts in technical interventions into the landscape of mass graves crucially redefined the value of such spaces. In sieving the soil, they erased some whilst preserving other traces of the recent past. The blue overalls, which I wore in common with everyone else on the site, made me feel clinical myself, as if tasked to make the graves salubrious again. Turning over earth, measuring bones and collecting corpses, site workers' processes of sense-making in effect turned pathology into physiology, construing remains as evidence, and the past as an instrumentality serving future uses.

But that night, hours after taking a shower, and returning my blue overalls to the team, I wondered how I would have read that very landscape, if it had not been the site of an intervention. How would I have understood that day if I had not been accompanied by forensically educated scientists and technicians? What is the meaning anyway of a sloping mint field, when not marked off by red flags and a yellow ribbon? To the inexpert eye, it tells nothing. The field was like any other scented field in summer; only the bodies interred in it made it a crime scene. The supposed, then confirmed, presence of bodies had played as much into my conception of the field as the forensic scientists', evaluating the land and enabling my assessment of their work. In order to restore a link between the temporality of an unthinkable past and the present, some form of intervention (comparable to the bodies' exhumation and examination) had been essential. At the same time, the connection between ancestors' graves and of those of contemporary war victims was irresistible; it was impossible for science, in whatever name, to divest itself of the resonances of metaphor.

Remains as Sites of Truth

Ever since the first positive results of the ICMP's DNA identification technique were publicly announced (on 16 November, 2001), DNA has been described in official reports and public documents as irrefutable evidence of a victim's identity. A population-based, DNA-led system of identification, based on profiling blood samples from close relatives of the missing, and matching them with bone samples from exhumed mortal remains, has speeded up the determination and repatriation of body parts/corpses to a very marked degree. In the last six years, ICMP experts have collected over 70,00 blood samples, identified more than 9,000 missing individuals across the former Yugoslavia, and developed an internationally acclaimed DNA-matching software tool.[22]

In the hope that a secure method of attribution would stimulate people to exchange information on the missing with greater speed and candour, the proponents of interstate reconciliation styled DNA in public discourse as the figurative basis on which people previously at loggerheads could be brought together. In the words of an ICMP web page, '[F]orensic Science in the Service of Truth and Justice' could potentially unite formerly warring parties. With time, the concept of DNA has taken root amongst, and gained the respect of, locals,[23] although it is by no means the case that the lay public understanding of DNA-based identification approximates to anything like the understanding of a trained scientist. Moreover, as Selma emphasized, it was precisely the DNA project that mobilized people in such great numbers around the cause of their missing. The ICMP's massive blood donation campaigns were not organized only in Bosnia and Herzegovina, Serbia, Macedonia and Kosovo, but through EU-funded outreach campaigns across Europe, the United States and Canada.

The effect of the ICMP brochures, public awareness campaigns, and, most of all, identification results, was to give rise to a notion that one could cope with the loss of loved ones, and attempt to secure justice on their behalf, by searching for them on a molecular level, through DNA profiles. While connoting personal and interpersonal reconciliation, the DNA-matching project also came, in ambiguous ways, to signify ideas of modernity, development and Europeanness. The restitution of an imaginary Yugoslavia criss-crossed by DNA collection and blood testing came to shape itself after a model of social and international cohesion between the states of the region and these entities in their political relationship to Europe. Further, the programme offered its participants a possible modality of 'biocitizenship', as they were mobilized in a civic function through acts of donating blood and offering information on their families. In the theorists' terms: 'They [biological citizens] are pioneering a new informed ethics of the self – a set of techniques for managing everyday life in relation to a condition, and in relation to expert knowledge' (Rose and Novas 2005: 450).

The concept of biological citizenship that Petryna (2002) introduced, and Rose and Novas (2005) further elaborated may indeed be useful when looking at the ICMP's Enlightenment ambition of promoting informed civil society discourse on the basis of DNA findings in postconflict Yugoslavia. Precisely such endeavours to educate the public about science and technology are discussed by Rose and Novas as prime aspects of the strategies for 'making up' a biological citizen. In referring to the 'making up' of citizens, they mean to designate 'the reshaping of the way in which persons are understood by authorities … be they political authorities, medical personnel, legal and penal professionals, potential employers, or insurance companies' (2005: 441). These broadly regulative procedures and modalities, for the authors, in turn entail 'citizen projects', or practices according to which 'authorities thought about (some) individuals as potential citizens, and the ways in which they tried to act upon them'(ibid.). To parse this in very compressed fashion, then, it is not categories of biology or the body that vest persons with 'biological citizenship'; rather, citizenship is also predicated on persons' and groups' ability to participate in social and public constructions (or contestations) of the biology of the body. For Rose and Novas, people's ability to engage with state and other institutions on the basis of an informed understanding of science works to relegitimate public provision (and, more widely, the processes through which citizens are constructed in relation to power).

In one sense, though, this optimistic rendition of biocitizenship fails to accord with my sense of local people's practice. As I have described above, many people I worked with at different times felt pressure (emotional and political) to give evidence as to the presumed state of their missing relatives. But they understood this need as answering to a familial imperative, rather than emerging out of (or being realized as) a form of civic duty or right. It is not clear that people would have come forward had they not been vexed as to their missing relatives' whereabouts; or whether they could have offered information merely on the basis of a systematic institutional intervention into (or redeployment of) their bodily potential, to adopt the Foucauldian language of biopolitics developed by Rose and Novas. People took a long time to utilize the ICMP frameworks as a means through which they might repatriate remains or claim certain benefits out of their situation *as a right*, on the basis of their being war victims or seekers after reconciliation. In other words, people claimed a form of emotional satisfaction, rather than their rights. For most of the time their emotions ran on fear and hope (that their missing ones would be found alive, or that, if they were indeed dead, their bodies would not have been desecrated, so that they would be able to bury them whole); these feelings, that is, were the wellspring of public participation in DNA citizenship programmes. These unsettled emotions, not the recognition of rights as such, shaped the form of their organizing:

> It was a nightmare. I was so afraid to think something had happened to him, that I didn't want to say aloud to myself, let alone to anyone else, that he was gone. I was frightened that just fearing it and thinking badly of it would make him dead. Imagine what would happen if my husband came home, after all, only to find that I had declared him dead and missing? That would be enough to kill him by itself! ... But he never came home. By the time I gathered up the courage to go and declare him missing, they [the ICMP] were already pretty good at identifying missing people via these DNA methods. So I gave my blood, our son gave his too, and my husband's sister and brother also gave blood. In the beginning I was afraid to ask them anything. I tried to contact them as little as possible, as I was afraid they had too many women calling them all the time, and I was afraid that they would tell me something horrible. I just waited. After two years of waiting, I have received a letter notifying me that the DNA identification of the commingled human remains in one of the pits they excavated a year ago, confirms that six bones in it were Zlatko's [her husband]. That meant that he was dead for sure. I was shattered. Completely. Particularly when they asked me whether I would take the 'identification'. I don't know how, but all of a sudden I got back all the courage I had lost, and I said no to them, that I had given them all of my husband and didn't want back only six bones. They were kind to me, they listened and then explained that there was only a very slight chance they would recover the whole body, that their funds were running low and they there was no way that they could run a DNA test on every bone they found – that would be far too expensive. But I could not accept that. So ... they are still searching for his body parts ... Sometimes, though, you know, I'm not sure, sometimes I think that I should have taken these six bones and just buried them. But then I do know, I feel, that I must bury Zlatko properly. To have six bones only is just not decent. It is not. (Ivanka, 44 years old)

Ivanka's courage in demanding that the ICMP carry on searching for her husband's remains is not only structured by the confirmation that her

husband really was killed, and that she could no longer *make him dead* by
'wrongly declaring his death'. Her access of courage, I believe, was also
predicated on changes in the social and political treatment of the war
bodies over the past few years in the former Yugoslavia – in the time, in
fact, during which she was waiting for her letter to arrive. I will try to
explain through the following example.

Almira hoped for years that she would hear from her father, who
disappeared in the early days of 1993. She felt particularly lonely and
desperate as she had also lost her mother at the age of three. But the
financial pressure of raising herself, her sister and younger brother, and
paying rent, forced her to barter the hope that he might be alive for a £220-
per-month pension. She gathered the paperwork and two witnesses and
declared him dead in 1999:

> Even though I was positive by then that he must have died, it felt such a
> loss, such a degradation to say that officially, you know, in public. I have
> never felt more humiliated then when I had to get together all the
> paperwork. But I had to have the documents, if we wanted to claim benefits
> and Ljubiša [her brother] receive his entitlement to a scholarship.

At the time of my fieldwork, the steps that anyone needed to go through
to claim and authenticate their missing relative's body were extremely
complicated. In much the same way as at the start of the war, people were
generally reluctant to declare their relatives missing, although by the end
of the 1990s, some certification of loss or absence had become essential to
many families' welfare. Women could not remarry if they were unable to
prove their late husband's death (since in law missing persons are
accounted neither alive nor dead); children could not ask for pensions
and other financial entitlements in the name of their parents; and many
people's living conditions, such as the roofs of their houses (if still
standing), water and gas mains could only be repaired on the basis of
funding attached to specific types of documentation. People who were
dislocated, who had lost everything during the war, stood in especial need
of official texts confirming their hardship and identity. Very often, these
two were the same thing; people had nothing to show for themselves but
penury.

This situation changed, however, when in November 2004, the
Parliamentary Assembly of BiH adopted a State Law on Missing Persons,
one of the first of its kind globally, enshrining the right of all families to
know the fate and whereabouts of their missing loved ones. To ensure this
right, the law stipulated a mechanism (the Missing Persons Identity
Protocol) according to which surviving family members could register
non-returnees with the Commission on Missing Persons (*Institut za nestale*

osobe, BiH) through the ICMP. Additionally, the law laid out principles for improving the search process for missing people, established a central database, coherently defined a 'missing person' for the first time, and provided for some of the social and other rights of the families of missing persons. These measures relieved some of my respondents[24] of the need to declare their relative's death as a precondition of access to the family apartment or house.[25] The law also filled in the gaps in international law through tasking the state with searching for remains and returning them to families. Further, families not covered by pensions, veterans' benefits, or benefits for civilian victims could sign up for missing persons benefits.

There is a difference, however, between the ability to claim rights and a guarantee to these rights. Many of the families I spoke to doubted that the law would allow them to see any reparations soon. As one man stated: 'It [the law] would be a good foundation if there were any money, but there isn't any' (Esad, 48 years old). In theory, the state missing-persons fund in Bosnia and Herzegovina should have been paid for by the three internal governments created after the war – the Bosniak-Croat Federation, the Serb Republic, and the Brčko District. But a general shortage meant that no one invested substantial amounts into the funds. For people to grasp reclamation as a rational process, there has to be a perception of a responsible subject (the state, an institution, or the individual) as being willing to consent to and implement (in this case, to finance) the positive outcome of claimants' cases.

I have described people's feelings of embarrassment and humiliation in the face of pursuing claims not simply to suggest their moral sensitivities or hardiness, but because such feelings capture something of the interpenetration of local and international impulses and imperatives in postconflict Yugoslavia. Both the idea and practice of claiming in respect to war losses was heavily reinforced by the presence of international bodies in Serbia and Bosnia-Herzegovina. A discourse and set of practices disposing of rights-concepts in relation to dead bodies that existed before the war has become increasingly prevalent in relation to the missing bodies of those lost in the war; further, ideas as to the information which bodies may yield have been decisively shaped by the technologies and ideology of the institutions, such as the ICMP, through which people process their claims for recompense and the 'right to know'.

Human Remains in Praxiography

In her book of that title, Mol ethnographically explores 'the body multiple' (2002) 'and its diseases in all their fleshiness', investigating the ways in which tensions between sources of knowledge and styles of knowing are negotiated in present-day allopathic medicine. If instead of bracketing the practices by which objects are handled, we foreground them, she suggests, reality multiplies (Mol 2002: 5). Further on, she reminds readers of the disciplinary-historical processes through which medical anthropology founded its subject. Social science's first step in the field of medicine was to define illness as an important conceptual object which could be held apart in a certain sense from disease's mere physicalities. Next, anthropology stressed that whatever doctors said about 'disease' represented a narrative or form of discourse, and as such partook of a realm of meaning, needing to be interpreted in relation to the specific perspective of the speaker. Mol proposes a further third step as formative of the methodology of a social science of the body – to foreground the practicalities, materialities and events according to which 'disease' ceases to be entirely pregiven as an ontology, but becomes a part of what is done in practice (Mol 2002: 12). If one is to understand how disease is 'being done', then the materialist turn to the physicalities of living, diagnosis and intervention is a necessary one. The researcher is enjoined to study a praxiography of instances as these fail to be entirely subsumed, on the one hand, by the sheerly bodily, and, on the other, by the psychosocial matters.

Projecting this logic onto the events I witnessed during my time in the former Yugoslavia, the thought occurred that the body of evidence collected on battlefields, in mass graveyards, DNA test laboratories and in courts potentially formed an object of study illuminating the *diseases of the communities* to whom they pertained. The language in which bellicose emotions were expressed during the war, and the idioms of subsequent traumas, losses, claims and eschatologies of the dead all correlated to, and spoke of, *communities' illnesses*. Mol's formulation, further, of discrepancies between 'sources of knowledge and styles of knowing' seemed very adequately to point up the slippages between the explanatory models offered by war victims, the relatives of missing persons, the recognized perpetrators of atrocities, forensic professionals, lawyers and politicians. Supposing, then, that Mol's framework offers a basis for a productive reorientation of my enquiry's concerns, it would seem to follow that, instead of seeking to extract some *cognitive* benefit out of my analysis of the Commission's and others' intervention into remains, I should shift, rather, from an epistemological to a praxiographic analysis of events.[26] Much of the rest of this chapter, then, deals in even more close-up fashion with the physical reality of bodies and body parts.[27]

Retrospectively, too, I understood that my research had wedded itself to a praxiographic method (long before I had read Mol's monograph), since it was the only one practical in the face of people's sensitivities about discussing human remains. People would either clam up when I raised the question about the traffic in body parts, or (as I suspected) elaborate or fabricate; others took the topic for granted, seeming thereby tacitly to project their own ideas about bodies and parts onto my supposed knowledge. Although people were encouraged to vocalize their experiences under a motto of national reconciliation, many I met did not want to, or could not, talk about the events of the past ten years. People's 'respectful silence' on the subject of war bodies, meanwhile, treated them in an illocutionary way: the communicative effect of their non-utterance and allusions underlined the importance of the topic, while forbearing to disclose itself in explicit statement. This meant that the knowledge I could more easily access was that enacted in daily events and activities, and located in laboratories, missing relatives' ICMP forms, DNA-extracting procedures, reparation claims' drafts, and so on – that is, within what have appeared as 'the full materiality and phenomenality of experience' (Mol 2002: 32).

More importantly, while the identity of those searching for proofs of a connection between themselves and those that were buried in mass grave-sites could only be *expressed*, the identity of those interred, on the other hand, could only be *performed*. This identity could only be stated within an order of procedure. First, bodies had to be identified. A dead body did not express anything on its own, but rather practices of intervention *performed* its reality.

And the mass graves were indeed a reality (whereas the causes, reasons and justifications for the war atrocities often appeared as narratives only). But it was a reality often unintelligible to me. The only way of making sense of it, and not mistaking a crime scene for a mint field, was through appropriating a certain technical knowledge.

Anatomy, for example, turned out to be a very helpful idiom for talking about past atrocities. In the ICMP's reassociation rooms, spaces for body bags full of commingled remains (too great in number to be DNA tested), experts devoted themselves to the knowledge work of turning bodies into objects. The same skulls, pelvic bones and jaws that I saw scattered in the mass graves seemed even more startling once respectfully and carefully set up in rows. These rooms lined up parts in their hundreds, even thousands; and the only way for me to decipher the scene was to treat it as a puzzle. Only someone with a knowledge of anatomy, it seemed, could reassemble the body part pieces into the requisite number of skeletons. But in these rooms what first appeared as a strange, respectful form of civic organization turned out to be a practical manoeuvre. The remains had to be lined up neatly before they could be put together. Mended and patched, skeletons gained a different status, however. They started to represent someone. The practices that 'performed' human remains' identity turned corpses or piles of body parts into objects, and then objects into individual war victims. The painful present, as exemplified in human remains, was articulated again into the temporal sequence of the past, enabling skeletons to retrieve their names, surnames, vocations, lives. In consequence, the identification of body parts offered the possibility of future commemoration. Whatever the outcome, though, the reconstruction of physical individuals proceeded through strictly technical processes. Technicians enact victims' individualities by measuring their hips, enclasping their dental remains, recognizing and naming their body fractures. These processes of diagnosing and reinstilling bodies' personality also inextricably and ineluctably construct dead persons as the evidence of crime, and as missing ancestral links.

Apart from allowing us to describe how things are being done in practice, such a 'praxiographic appreciation of reality' (ibid.) may also give a further push to analysis. Not only do practices fundamentally intervene into their objects, forensic methods of identification seem predicated on effecting some sort of change in the object with which they interact.[28] Classification and reconstruction alter objects. Moreover, alteration was not only the effect of the practice but in most cases represented the rationale for its performance. The graves, landscapes and ICMP waiting rooms had to be reshaped, almost renewed. Practices had to yield a moral transformation. Producing such an effect (and affect), interventions could not be understood simply as a way of organizing individual lives (of people and corpses), but set out to shape postconflict societies as a whole.

At the same time, it is not only processes which alter these objects or people's relations to them. The land in which remains had rested had also altered or, more precisely, eroded them (at least partially). But interestingly, even decay became valuable in processes of reclamation, insofar as it rendered reconstruction possible. The postmortem dissolution of the body,[29] and the postmortem history of bones, expressed the force of an interplay between opposing agencies of preservation and destruction, in the sense of cultural memory as much as of soil erosion.

Conclusion

During the Yugoslav wars, the body was often invoked as a metaphor for national wholeness and resilience; it symbolized people's claims to a particular territory and motivated them to take sides or take up arms. In the aftermath, when many bodies had become body parts, human remains were in turn invested with a unique significance, as carriers of certain defining forms of identity (whether as victim, perpetrator or national subject), and as vehicles through which particular versions of modernity and reconciliation could be legitimated.

If in wartime political rhetoric and practice the body of the Serbian people was reanimated in battle, it appears that in medical and peacetime rhetoric, Serbia becomes whole again through the rightful assignation of the bones of its war-missing. DNA testings and reassociation techniques apparently yield a new solution to questions of conflict through a recourse to scientific incontrovertibility. Genetic accreditation, along with the reasocciation practices, was offered as a definitive and scientifically sanctioned verdict on a person's victim identity. Moreover, in applying certified expertises (of archaeology, physical anthropology, pathology, physiochemistry, odontology, etc.) to a local setting, in an attempt to solve crimes and restore identity to the missing, forensic researchers suppose themselves to be contributing to both individual and political (or inter-state) healing. Forensic identification enables body parts to accede to a metonymical status not merely in their capacity as evidence of a greater crime, but symbolically, as tokens of the ways of life, communities or nations that were violated during the war. The signs left on the body – by the perpetrators, by scientists and by time – assume different values in different contexts, but would all seem to be construed according to an evidentiary function. Bodies are evaluated; they certify and attest. Hence victimhood, citizenship and modernity may all be predicated on DNA methods of identification. Moreover, it is no longer possible to sustain the claim that the private treatment of the remains, i.e. supposed barter, represents an index of 'backwardness and atavism', as it was in the early 1990s. On the contrary, the foreign agencies today see the active involvement of locals in questions of the remains and their commemoration as 'crucial, progressive, liberal, and moral'.

But *does* restoring the remains promote reconciliation? How effective, painless or speedy a process is this? And how hygienic can the notion of reconciliation be, if presided over by the image (however attenuated by scientific methods) of dismembered corpses, projected onto a visually distorted symbolic register, as this continues to dominate people's emotional landscape? The projective processes of rememoration, even if intimately connected with mourning, in my view have both affirmative and negative aspects: affirmative, because they can bring families some sense of resolution, and negative, because they are liable to manipulation in the context of revanchist and reactionary politics. I have recorded many cases where the technical identification of the remains has functioned also as a process prompting the resuscitation of illusory or abusive images of national wholeness. Now that the DNA test has been established as a supreme proof of a victim's identity, the dead body's biology has became a supreme proof of ethical belonging. As a seventeen- year-old schoolboy, the greater part of whose family was slaughtered in Srebrenica, asserted:

> I did give my blood for the ICMP's DNA testing. Most of the people I know did that. And that's right. Only those who were DNA-tested can prove that their families are not guilty [of war massacres]. This is why only DNA-tested people, like us in Bosnia, should get into the European Union.

Among many of my respondents, the scientific attribution of identity through DNA methods had buttressed popular mythologies of underlying physical differences between Serbs, Croats and Bosniaks. Further, more far-reaching and complex political questions, like membership of the EU, were referred to the tests. And the EU-funded campaign for collecting DNA samples overseas was read by some as a drive potentially rallying a genetically coherent Serb (or Croatian or Bosniak) population from around the globe – thereby lending itself to political claims for the self-sameness or homogeneity of these peoples.

Many relatives themselves, though, want rather to demystify the bones that have been sent back to them – to divest them both of the juridical, scientific and proto-civic narratives in which they have been wrapped, and also of the cargo of unpleasant historical associations they have been forced to carry:

> That wretched shambles that happened fifty years ago, that happened again in the awful shambles of ten years ago. The same bloodbaths, and the same excuses. I am so tired of it all ... I am tired even of talking about it. I want to change the topic away from the war – let it die once and for all. I feel like a stranger here. And that's horrendous. Believe me, it's far easier to be a stranger in a foreign land than in your own. (Višnja)

Expressing sentiments like these, some have set their faces against the conscription of the remains into any project of reconciliation or European integration. Were the remains to be used as tokens of, say, Serbia's willingness to face up to its guilt, they feel that their pride, or their respect for their ancestors or recent dead, would be effectively demeaned. While the pursuit of accountability through human remains is understood as instrumental and potentially valuable in legitimizing democratic government, repairing bridges with the international community, and restoring dignity to individual victims and their families, the politicization of the bones continues to arouse a disquiet about atavism. People hold to an idea of modernity while still expressing the hope that the 'deceased will rest in peace' (Ružica), that the missing will be free of the interference of the ICMP, which only discomfits them. The healing sought by families is not necessarily a quick or public process.

Rather than making an unproblematic distinction, then, between local desires and procedures to make sense out of dead bodies, on the one hand, and the 'proper' or hygienized objectives of transnational institutions, on the other, this analysis reads both parties as caught within the same rhetorics, because both centrally deploy figures of the restored body in their imaginary projection of some shattered whole (families in the case of relatives, and the nation or international harmony in the case of agencies). It is only through making claims for certain sorts of identity and structures of belonging that any agent is able to articulate their goals in the supposed process of national healing.

In a way, both groups treat human remains in an evidentiary or immediately compelling mode, as if no further interpretive work is required on whatever scientists discovered in the mint fields. But it seems imperative to insist that scientific attribution is only the first step in a complicated and multi-sided process of reckoning with the past, that will involve families as well as institutions assimilating individuals' fate into possibly many narratives of justification, rationalization, culpability and forgiveness. When the Second World War Yugoslav graves were disinterred, the political context in which the bones were used (or abused) prior to the 1990s wars became, rather, less perspicuous; bones do not interpret themselves, nor is the meaning of evidence always evident. In the same way, it seems hasty and wishful to suppose that the retrieval of war bodies now will unproblematically serve the cause of national reconciliation. The political context in which the actions of attribution and return will be understood has rather still to be specified, according to whatever social trends most successfully capture the demands and aspirations of Serbs and of the international community at large.

I have argued that postconflict attempts at sense-making should be understood not only as relating to the historical legacies from which they emerge, but also as contemporary practices finding their effectivity in the present. The contemporary intervention into the land and the bodies it holds defines the past. In order to see a mass grave rather than a field of mint, certain practices have to be granted, or to be allowed to take hold. If nothing else, a yellow ribbon has to encircle a grave. To show respect for the dead, a mother must sit by her son's tomb, even if it is empty. A wife must persuade herself that declaring her husband missing may precipitate his death. Given how raw these private sense-making practices may feel, larger-scale practices of collecting, measuring and analysing bodies may appear invasive; when, however they are recognized as participating in the same processes of grief and accounting as more private acts, they are more usually welcomed.

Acknowledgements

My sincere acknowledgements go to Marilyn Strathern, Helen Lambert, Maryon McDonald and the anonymous reader for their comments on the original version of this paper. To all these people who, during the time of my fieldwork, extended their hospitality, and shared their time and experiences, my most grateful thanks. This work would not have been possible, had I not been supported by the Wenner-Gren Foundation with the Individual Research Grant, by the Ad Futura and by the Royal Anthropological Institute with the Sutasoma Award. Parts of the fifth section of this paper, in its earliest version, were published in *Cambridge Anthropology* 25(3): 61–71, 2005/2006.

Notes

1. My Serbian fieldwork was based in the capital, Belgrade, on account of its centrality to the research both as the home of local and international bureaucracies dealing with human remains and as the symbolic heart of the country's relations of 'national kinship'. However, throughout my research, I ranged widely across the former Yugoslavia following those professionals and institution staff searching for remains and processing parts' return to families.
2. See Čajkanović (1994 [1910–24]); Kulišić (1979); Lowmniasky (1996 [1979]); Zečević (1982); Čolović (1984); Dvornić (1994); Petrović S. (1995); Bandić (1997); Jovanović (2002). Similar rituals have been observed in the other parts of the world (see Hertz 1960 [1904–6]; Ephirim-Donkor 1997; or Swain and Trompf 1995: 157).
3. A specific concern with the exhumation of bones (*kosti*) once they were free of the flesh was especially important insofar as bones were understood to bind the soul to the profane world. In Russian Orthodox doctrine, a dead person is revealed as a saint not only through miracles but also because his corpse does not putrefy (see Verdery 1999). In Serbia, on the contrary, the body that does not decompose represents a proof that the deceased was a 'bad man', leaving open the further possibility that s/he might turn into a vampire (Jovanović 2002: 131).
4. Considering that Orthodox Christian doctrine teaches that the human being consists of a body (*telo*) and soul (*duša*), both of which are believed to be resurrected at the Second Coming of Christ, the natural decomposition of the body is key to the traditional prohibition of both embalming and cremation.
5. To understand the role that Kosovo plays, both as historical reality and as metaphor, in the constitution of Serbian cultural and national identity, see an outstanding article by Bakic-Hayden (2004). To dismiss the Kosovo theme as something solely reserved for fictive 'representation' (epic and myth) and isolated from 'fact' (history), Bakic-Hayden explains, would be to

underestimate the degree to which the popular conception of Kosovo has the power to mobilize certain actions and identifications in Serbia. Specifically, the myth of Kosovo served to mobilize terrestrial claims over other regions in the former Yugoslavia, rather than only over Kosovo proper.

6. According to a traditional custom prescribed by the Christian Orthodox Church, the remains of famous people and saints, after their traditional excavation and washing, are ritualistically carried around Orthodox monasteries.

7. For this also see Verdery (1999: 18), and her simile with the travels of Frederick the Great's human remains around the boundaries of present-day Germany after its reunification.

8. There is an extensive scholarship on the 'folk religion of Serbs' (*ljudska religija Srba; narodna religija*) specifically by Kulišić (1979); Zečević (1982); Čajkanović 1994 [1910]; Bandić 1997, and others. Although Serbs accepted Christianity as early as the ninth century, Bandić argues that people often understood and practised Christianity in non-Christian, pagan and animistic ways. God often had and still retains the characteristics of a pagan deity, and beliefs in magic continue to complement religious celebrations. In this sense, popular or pagan beliefs still play a central role in establishing and maintaining the system of ritualized rules governing the specific procedures of bidding farewell to the deceased (cf. Bandić 1997).

9. Bandić (1997) argues that it is more correct to speak of the 'popular or folk religion of [the] Serbs' rather than of 'Serbian folk religion', since little firmly distinguishes the Serbian set of religious practices from those performed by other Orthodox Slavs under the jurisdiction of the Church. For Bandić, specifically Serbian phenomena only denote '*krstna slava*' and '*svetosavski kult*', leading to his suggestion that the pagan heritage and magical animistic religious beliefs of the Serbs rather represent part of a shared Slavic patrimony.

10. These forms are regulated by the popular belief that one year after death, the soul of the deceased joins the souls of the ancestors in a joint cultus. The transformed soul of the late person attains the status of a distant dead person, or common ancestral spirit, functioning as a protector of collectivity and symbol of collective identity. After a year, when the soul is taken finally to part company with the properly buried body of the deceased, it begins a search for a new home in which to live. Souls may go into tombstones, ledgers, into the trees (the reason that many people plant fruits by the graveyards), or into animals (so-called *senovite životinje* or *psychopomps*, such as wolves, snakes, cockerel, pigeons, etc.) There is a recorded belief that impure souls go to live in mice; finally, souls can also migrate into other people. Ancestors usually communicate with the living in the image of a man, a stranger or beggar. The belief that souls can reincarnate into stones is also found in Israel (Weingrod 2002).

11. '*Da bi mu barem duša našla spokoj*'.

12. This ritual is still performed in some areas, including rural Greece (see Danforth and Tsiaris 1982; Seremetakis 1991; Verdery 1999: 45), Romania and

among Orthodox Albanians (Đorđević 2004 [1923]). Importantly, although the belief is associated with Orthodox Christians, the expression of concerns with the proper disposal of remains, signally bones, is also widespread among the Catholic and Muslim inhabitants of the former Yugoslavia (cf. Bringa 1995; Ballinger 2003).

13. A number of my respondents complained that one of the most vexatious features of Tito's suppression of the past was their being frustrated in any search for the graves of their beloved, meaning that they were prevented from formally commemorating their deaths. In his solidification of a 'Yugoslav identity', Tito moved to forestall any airing of different nationalities' grievances. As people were prohibited from publicly naming or criticizing perpetrators, they keenly felt that 'the souls of their dead were still tormented'. Another group of informants, whose relatives were purportedly not victims of the regime, offered a rather different, more favourable account, suggesting that although Tito was an autocrat, he made life in Yugoslavia pleasant, secure and affluent. The argument that totalitarian leaders, who suppressed open disagreement over the conflicted past, imposed clarity instead of chaos, is also one found in Green's ethnography of Epirus, in northwestern Greece (see Green 2005: 64). Yet a third group amongst my interlocutors stated how much they hated 'the patronising analysis that people in the former Yugoslavia lived under the Iron Curtain and wanted to be subdued only for the sake of clarity and order' (Goran, 64 years old). However, many older people with whom I spent time stressed their love for *Tito's socialism*, [saying] how proud they were 'to have lived through it'. Tito's period and policy, according to the opinion of a large number of people, 'was a very nice period, which will not be repeated', and in general, people regret the end of his rule and the harmony between ethnicities he sustained after the extraordinary bloodletting of the Second World War. Once President-for-life Tito was gone, wartime and earlier grievances, it is widely accepted, were exploited by politicians jockeying for power.

14. Although Yugoslavia was under socialist rule under Tito from the late 1940s until the 1980s, when religious practices had to be, at least publicly, quelled, an intense burial regime was still encouraged and regulated by the state. In former Yugoslavia everybody had to pay rent on the grave-sites of kin buried in the cemeteries. The fee was, and still is due to the state, and if it goes unpaid, the burial site is leased to someone else (with the contents removed).

15. 'National kinship' is a phrase that I have provisionally coined for the purposes of conveying my assumption that in former Yugoslavia, along with kinship relations, understood as personal and domestic kin networks, one may talk of national kinship networks and alliances on a political level. The idea of kinship on the national level – for example, the treatment of all Serbs as kinsmen on the basis of their shared ethnicity – propagated a national value system that functioned as a kinship value system, encouraging respect, loyalty and solidarity amongst Serbs in regard to the fighting army, politicians and the state. Pro-war politicians in Serbia, as well as those in Croatia, nurtured these ideas in order to fulfil specific political imperatives during the war.

16. Notably, the ICMP was (or is) the first organization of its kind to be created specifically envisioning postconflict situations; its remit of excavating, identifying and returning such an enormous quantity of missing persons is also unprecedented. Today, this organization and a few others with similar missions are trying to set up work in Iraq, and adjusting their logistical set-ups to future possible postconflict settings.

17. The idea that formerly disempowered, oppressed or aggressive societies or countries can achieve modernization, and the recognition of the international community, through a public acknowledgement of truth, i.e. their wrongs (as, for instance, accounted for through Truth and Reconciliation Commissions) has been a feature of international relations discourses from the early 1980s and is especially salient in the Serbian context (see Petrović 2003).

18. Though it was always apparent to the local population that the ICRC (International Committee of the Red Cross), and later ICMP, was building up a file on missing persons, many locals took their time to come forward and report their losses officially (and some have resisted doing so). People feared that in reporting missing peoples' information, they would indirectly be informing on their loved ones' activities in the war; others held to the hope that their missing relatives would soon return anyway; and most of the women I spoke to confessed how much it confused and distressed them even to imagine that their husbands, sons, brothers or uncles would not return. Reporting that someone was missing was tantamount to accepting their death, which many people were reluctant to do.

19. The ICMP, the organization widely discussed in this paper, was not the only organization that endeavoured to secure the cooperation of governments from those formerly hostile countries of Yugoslavia, with the end of locating and identifying persons missing as a result of armed conflicts. The International Committee of The Red Cross (ICRC) and later the Missing Persons Institute (*Institut za nestale osobe Bosne i Hercegovine*) were and still are trying to encourage public involvement in their activities as these are billed as contributing to 'the development of appropriate expressions of commemoration and tribute[s] to the missing' (as stated in the ICMP mission).

20. The *tespih* is the traditional Muslim rosary, said to be imbued with the spiritual power of an endless number of prayers intoned by members of one's house and community (cf. Bringa 1995: 159).

21. As essential articles of sociability, clothes are integral to the revivification and rearticulation of the personhood of dead persons (cf. Mol 2002). It would always evoke strong emotions when a piece of clothing was found amidst piles of remains in a mass grave. It was somehow easier to look at dead people's bones than their clothes. Bones could possibly be mistaken for animal remains or for the deposits of an ancient past, while there was nothing 'archaeological' about clothes – they were violently contemporary. Many of the victims' clothes were still bright and their textures still fine or raised.

22. The number of DNA matches for missing individuals from the former Yugoslavia as of 17 February 2006 stands at 9,220. The total number of blood samples collected and processed to obtain DNA profiles, which are

subsequently located onto the ICMP DNA database, is 78,559. The total number of individual cases of missing persons represented by the samples collected is 26,740, and the total number of bone samples from which the ICMP has successfully obtained DNA profiles has reached 17,331. Source: ICMP official website http://www.ic-mp.org

23. My fieldwork threw up many instances where people who wanted to find their missing did not ask too many questions about the processes which would conduce to their doing so. Nevertheless, these people gave vivid and emotionally engaged accounts of the practices of the blood giving. Many aired suspicions: 'Why do they need our blood after all?', 'Will they sell it?', 'This sounds to me like another vampire story', etc.

24. This law, importantly, pertains only to the citizens of BiH. As the law is the first of its kind in the world, many speculate that it might prove a model for other countries. It has been unable, however, to accommodate other 'seekers after rights' in the former Yugoslavia. The press reports that family associations in Kosovo, in Serbia, as well as Iraqis and those affected by the Asian tsunami, who have to deal with the missing, are keen to put a similar law in place. In this they are especially seeking to access the unique expertise of the ICMP in extracting DNA profiles from bones and teeth, which enables it to identify victims of natural as well as man-made catastrophes, such as the 2004 Asian tsunami and Hurricane Katrina, in 2005.

25. Under the new law, relatives can sign up for what is called a 'temporary trusteeship' of the property.

26. Some would claim that Mol's analysis replicates presuppositions common to poststructuralist anthropology in general. Her analysis, however, has decisively influenced my understanding of the war-harmed bodies, and I want to acknowledge that. Importantly, although Mol's work has enabled me to conceive of the multipleness of the body-parts phenomenon in postconflict Serbia, my reading of her text was not stimulated by the evident fact that the bodies I was researching were usually fragmented or scattered. There is more to her figure of the 'multiple body', that is, than the physical fact of bodies' diasporas. In other words, the multiple body does not stand for a fragmented or plural one.

27. The categories with which I operate here are ethnographic and not analytical.

28. During the Yugoslav wars international political and military interventions aimed to put an end to the fighting. The intervention of laboratory technicians as they filled their test tubes with solvents and reagents served to translate one kind of knowledge into another – a blood test into a DNA proof of victimhood. I have myself intervened into the landscapes of mass graves whilst helping technicians record the quantity and state of remains found.

29. But then one could interpret the decaying body as a process, too, and not only as an object.

References

Bakić-Hayden, Milica. 2004. 'National Memory as Narrative Memory. The Case of Kosovo', in Maria Todorova (ed.), *Balkan Identities. Nation and Memory*. London: Hurst & Company.

Ballinger, Pamela. 2003. *History in Exile: Memory and Identity at the Borders of the Balkans*. Princeton, NJ and Oxford: Princeton University Press.

Bandić, Dušan. 1997. *Carstvo zemaljsko i carstvo nebesko*. Beograd: Biblioteka XX vek.

Bjelić, Dušan and Obrad Savić. 2003. *Balkan as Metaphor: Between Globalization and Fragmentation*. Cambridge, MA: MIT Press.

Bringa, Tone. 1995. *Being Muslim the Bosnian Way: Identity and Community in a Central Bosnian Village*. Princeton, NJ: Princeton University Press.

Butler, Judith. 2000. *Antigone's Claim: Kinship Between Life and Death*. New York: Columbia University Press.

Čajkanović, Veselin. [1910–24] 1994. *Studije iz Srpske religije i folklora 1925–1942*. Beograd: Srpska Književna Zadruga, BIGZ, Prosvet.

Čolović, Ivan. 1984. *Divlja književnost*. Beograd: Nolit.

Danforth, Loring and Alexander Tsiaras. 1982. *The Death Rituals of Rural Greece*. Princeton, NJ: Princeteon University Press.

Denitch, Bogdan. 1994. *Ethnic Nationalism: The Tragic Death of Yugoslavia*. Minneapolis: University of Minnesota Press.

Dvornić, Milan. 1994. 'Pogrebni običaji baranjskih Srba', *Raskovnik* 75–76: 26–32.

Đorđević, Tihomir. [1923] 2004. *Naš narodni život*. Beograd: Prosveta.

Ephirim-Donkor, Anthony. 1997. *African Spirituality: On Becoming Ancestors*. Asmara (Eritrea): Africa World Press.

Green, F. Sarah. 2005. *Notes from the Balkans. Locating Marginality and Ambiguity on the Greek–Albanian Border*. Princeton, NJ: Princeton University Press.

Hayden, Robert M. 1994. 'Recounting the Dead: The Discovery and Redefinition of Wartime Massacres in Late- and Post-Communist Yugoslavia', in Rubie S. Watson (ed.), *Memory, History, and Opposition Under State Socialism*. Santa Fe, NM: School of American Research Press.

Hertz, Robert. [1904–1906] 1960. *Death and the Right Hand*. Aberdeen : Cohen & West.

Jovanović, Bojan. 2002. *Srpska knjiga mrtvih* [The Serbian Book of the Dead]. Novi Sad: Enciklopedia Serbica, Prometej.

Kulišić, Špiro. 1979. *Stara slovenska religija u svjetlu novijih istraživanja posebno balkanoloških*. Sarajevo: Akademija nauka i umjetnosti Bosne i Hercegovine.

Lock, Margaret. 2002. *Twice Dead: Organ Transplants and the Reinvention of Death*. Berkeley: University of California Press.

Mol, Annemarie. 2002. *The Body Multiple: Ontology in Medical Practice*. Durham, NC and London: Duke University Press.

Lowmianski, Henryk. [1979] 1996. *Religija Slovena*. Beograd: Biblioteka XX vek.

Petrović, Maja. 2003. *The Practices of Justice and Understandings of Truth: Truth and Reconciliation Commissions*. Available at: http://www.eurozine.com/article/2003-12-02-petrovic-en.html

Petrović, Sreten. 1995. *Mitologija, kultura, civilizacija*. Beograd: Salus.

Petryna, Adriana. 2002. *Life Exposed: Biological Citizens after Chernobyl*. Princeton, NJ: Princeton University Press.

Rose, Nikolas and Carlos Novas. 2005. 'Biological Citizenship', in Aihwa Ong and Stephen J. Collier (eds), *Global Assemblages: Technology, Politics, and Ethics as Anthropological Problems*. London: Blackwell.

Scheper-Hughes, Nancy. 2000. 'The Global Traffic on Organs', *Current Anthropology* 41: 191–224.

Scheper-Hughes, Nancy and Loïc Wacquant. 2003. *Commodifying Bodies*. London: Sage.

Seremetakis, C. Nadia. 1991. *The Last Word: Women, Death, and Divination in Inner Mani*. Chicago: University of Chicago Press.

Swain, Tony and Gary Trompf. 1995. *The Religions of Oceania*. New York: Library of Religious Beliefs and Practices, Routledge.

Verdery, Katherine. 1999. *The Political Lives of Dead Bodies: Reburial and Postsocialist Change*. New York: Columbia University Press.

Weingrod, Alex. 2002. 'Dry Bones: Nationalism and Symbolism in Contemporary Israel', in Jonathan Benthall (ed.), *The Best of Anthropology Today*. London: Routledge.

Zečević, Slobodan. 1982. *Kult mrtvih kod Srba*. Beograd: Biblioteka Susretanja.

Chapter 3

ON THE TREATMENT OF DEAD ENEMIES: INDIGENOUS HUMAN REMAINS IN BRITAIN IN THE EARLY TWENTY-FIRST CENTURY

Laura Peers

As Curator for the Americas Collections in the Pitt Rivers Museum at Oxford (UK), one of the tasks I have undertaken has been to update the most popular case in the museum, which contains the shrunken heads (*tsantsas*) from Shuar and Achuar peoples of South America, as well as scalps from Native American groups and Naga head-hunting trophies. The case is labelled 'Treatment of Dead Enemies' and provides examples of the ways in which dead bodies can be deliberately mutilated, dismembered, displayed, used to humiliate enemies, or to appropriate their powers. My work on the case has included inserting an illustration taken from a 1605 broadsheet, showing the severed heads of Guy Fawkes and his co-conspirators on pikes, to convey the idea that it wasn't just exotic 'others' who did things to enemy heads. I have also written new text for the case pointing out that in one form or another, such mortifications of the bodies of enemies are sanctioned in many societies: these forms of violence might be seen as a way of maintaining social order.[1]

This work has been done in a climate of sensitivity in Britain about human remains. Several groups have been appointed by government in recent years to consider the treatment of human remains, including the Retained Organs Commission, which examined issues connected with the retention of post-1948 medical specimens in England; the Church Archaeology Human Remains Working Group, which focused on burials from Christian contexts in England dating from the seventh to the nineteenth century AD; and the Working Group on Human Remains, of which I was a member,[2] which focused on overseas indigenous remains from AD 1500 to 1948. Interest in issues to do with human remains has also included debates over whether Dr von Hagen's 'Body Worlds' exhibition constituted educational or voyeuristic experience, and over whether human remains should be returned to indigenous peoples at the expense of scientific research. Human remains have been returned to indigenous groups from Exeter, Manchester, the Horniman and the Royal College of Surgeons, with more claims now being processed. The 2004 Human Tissue Act has insisted that institutions holding remains less than a century old be licensed to do so, and that they adhere to strict guidelines for the handling, storage and display of remains.

Human remains are equally sensitive at the Pitt Rivers Museum, where staff struggle to balance the Victorian displays with contemporary research and policies. Suggestions that the shrunken heads should be removed from display have met with approval by some staff and visitors, and by cries of 'Whitewash!' from others.[3] One visitor, who heard a rumour that the museum might remove the shrunken heads from display, was moved to write to us:

> I am appalled at this news as the heads are one of the most distinctive and interesting exhibits and never fail to interest friends of mine from this country and abroad. Despite the fact that the heads are body parts they are a fascinating and integral part of the museum as the public see it and [to] remove them would be an unforgivable act of censorship. It seems that political correctness had gone mad and could deprive us Oxford residents of a much loved part of the City's culture. (Pitt Rivers Museum visitor feedback form, 16/8/2003)

This visitor takes his own visitors to see the *tsantsas*; he admits that for him they are wrapped in social relations. He also admits a tension between their status as body parts (interpreted as parts of persons) and as parts of the museum (objects). And in his accusation of 'political correctness gone mad', he refers to tensions between social groups and the beliefs they hold about how one should relate properly to others. Just as these particular artefacts were profoundly social in Shuar and Achuar contexts, where they were created to steal souls from other groups for the benefit of one's own group, so they continue to be profoundly social in their British museum context.

Hallam and Hockey (2001: 43) have argued that the 'social lives of persons might persist beyond biological death, in the form of the material objects with which they are metaphorically or metonymically associated in social processes of memory making', and that social interaction with such material forms 'tends to destabilize subject/object boundaries such that material objects can become extensions of the body and therefore of personhood.' In this paper I go beyond their analysis to argue that human remains themselves are so wrapped in social meanings that they appear not only to act as extensions of persons, but as powerful social agents engaged in ongoing social relations: their social lives continue well beyond death. People learn to respond to human remains in certain ways and to participate in certain forms of social relations with the dead, just as we learn to respond to and have relations with the living. These patterns of behaviour vary within any society and in accordance with the definition of the dead. A surgeon, a coroner, a bereaved parent or widow, a minister or priest, and an atheist, for example, would all have different ways of thinking about dead kin or the dead they might encounter professionally.

Human remains held in scientific institutions and museums evoke this same gamut of social relations and meanings, from data to kin, from scientific specimens to persons. The sensitivities within Britain over human remains recently have seen significant shifts in attitudes and practice towards human remains, as articulated in new legislation and official professional policies for dealing with remains.[4] To a great extent, these shifts have been towards acknowledging the social identities and social nature of human remains, validating their lingering ties with the living, and encouraging forms of social interaction with the dead they represent – and with their descendants. In doing so, they both acknowledge and force changes in the relations of power between different social groups amongst the living· between the public and medical scientists, between indigenous people and former colonizers. 'What the bones provide,' as John Cove notes, 'is a locus for examining aspects of a more complex set of changing social, economic, and political interactions' (1995: 3).

I focus here on recent debates within Britain about human remains deriving from overseas indigenous peoples who were historically colonized by British and other nations: specifically, First Nations, Native American, Maori, Tasmanian and Australian Aboriginal groups dating from the period AD 1500 to 1948. While these are very different cultures with specific histories, they have shared experiences of being colonized, dispossessed, and subjected to forced assimilation and racism, and I thus gloss them together as 'indigenous peoples', bearing in mind the politics of colonialism and postcolonialism attached to that complex term.[5] These peoples have also shared similar historical experiences of having their dead removed by collectors for purposes of scientific analysis and representation. There are substantial collections of such remains in British museums: in a very partial survey of English and Welsh museums in 2002, sixty responding institutions held such material (Working Group on Human Remains Report 2003: 11). These particular remains have generated great controversy in British museum and scientific spheres recently, and thus offer a useful lens for understanding their different, intersecting sets of social meanings and the social relations they generate. I use the work of the DCMS (Department of Culture, Media and Sport) Working Group on Human Remains, and a certain volume of repatriation claims to Britain for remains from Maori, Australian Aboriginal, First Nations and Native American groups since 2000, as a specific focus for my analysis. However, this focus should be seen more broadly as characteristic of contemporary societies with historic overseas colonies and of postcolonial discourses in those settings generally.

This is, then, part of an ethnography of human remains within postcolonial contexts, as legacies of colonial relations which led to their collection and intellectual consumption within the home countries. While debates have also raged in North America and the Pacific across the 1980s and 1990s about indigenous human remains,[6] the nature of the debate is quite different in the UK. In settler countries, the debate was driven by the central presence of indigenous peoples, and focused on the poor social and political relations between living settler and indigenous peoples as much as on the disposition of human remains. In the UK, however, arguments have been characterized by the relative absence of overseas indigenous groups and have effectively consisted of different groups of British people speaking to each other, with very minor input from source communities.[7] In all countries, the debates have been (at one level) about what the appropriate social relations with the dead should be. Negotiating answers to this has in every case involved social change, and change in relations of power. In this paper, I explore the nature of such change within Britain, as seen through the lens of debates over the disposition of historic indigenous human remains.

While the paper is about human remains, rather than repatriation, repatriation will come into the analysis because it reveals and challenges deeply held beliefs and established patterns of power within and across societies: it is one of what Lambert and McDonald have called 'new types of transactions that focus squarely on the ... corporeal' which generate new forms of sociality (Introduction to this volume, p. 2). This is not just true for overseas material. The Alder Hey and Bristol hospital scandals (where in 1999 it was revealed that organs of dead children had been retained without parental consent for research) and subsequent public enquiries[8] also revealed tensions – problematic social relations of power – between scientists, medical professionals and the lay public in England, and could be regarded as an exercise in the repatriation of British human remains. As with other cases of repatriation, this process saw significant renegotiation of social and political relations amongst the groups involved: debates over who has authority over human remains become an important arena within which expectations about social relations with the dead, and about social relations amongst the living, are articulated.

Human Remains

Scientists, museum professionals and members of indigenous groups all speak with great certainty about the meanings of human remains. To some, remains equal 'data' to support research; to others, specimens, like rare documents in an archive, to be numbered and stored in compliance with the Human Tissue Act (HTA); to others, ancestors who need to be retrieved from enemy hands and laid to rest. Each set of meanings implies a different set of expected social relations and attendant behaviours towards the dead.

One of the central problems in trying to understand the social meanings and effects of human remains is that they are not only iconic, but ontologically unstable, and are apt to shift registers of meaning abruptly:[9] human remains lurch between signifying the living and the dead, person and object, specimens and ancestors. Katherine Verdery asserts that, 'among the most important properties of bodies, especially dead ones, is their ambiguity, multivocality, or polysemy. Remains ... do not have a single meaning but are open to many different readings' (Verdery 1999: 28).

Increasing the instability of the meaning of these remains is the fact that these different meanings are always co-present: human remains are always, for different sets of people, *both* data *and* ancestors, with the attendant expectations of social behaviour towards the remains implied by each set of meanings. It is the dynamic tension in such dissonance that gives human remains – and discourse around them – such force.

In considering the anthropology of the body in colonial America, Lindman and Tarter (2001: 2) have noted:

> Bodies are not only physical phenomena but surfaces of inscription, loci of control, and transmitters of culture. They are never unmediated; they are related but not reducible to cultural concepts of differentiation, identity, status, and power Encompassing both the physical and the symbolic, [the body] is enmeshed in the social relations of power.

Although Western scientific contexts have often appeared to strip human remains of social meaning, this redefinition itself is a social (and political) process. Only very particular dead, with very particular social positions vis-à-vis British society, had their heads stuck on pikes, were seen as desirable for inclusion in a collection, or wound up in the Treatment of Dead Enemies case at the Pitt Rivers Museum. Indeed, only very particular dead were available for collection. Looting of remains from graves was discouraged by the British Anatomy Act of 1832, but sanctioned in the colonies after that date to fulfil the desires of scientific research: dead bodies with one collective social identity ('primitive people') were deemed suitable by British people to collect, while the bodies of ordinary British people were illegal to collect – although those of the destitute, a socially stigmatized group, could legally be collected for dissection (Richardson 2001). In British museums and scientific institutions today, the remains of indigenous peoples from around the world exist in diverse forms which were created within webs of social meaning and relations in originating and collectors' contexts: rendered bones, excavated skeletons, shrunken heads, mummified foetuses and bodies, scalps on stretchers, hair specimens in envelopes, DNA caught in plaster live-casts, bits of tanned skin, and finger-bone necklaces. These, and other preserved forms of human cells, might be said to embody not only the social identities of the persons the remains derive from, but the social and political relations between one people and another, at the time of collection as well as today. In this respect, the varied and fragmentary forms of human remains from indigenous groups are no different from the whole corpses or retained organs involved in arguments over meaning within British medical contexts recently: entire or in pieces, in whatever state or form of preservation, human remains constitute an intense focus for the complex social responses that people everywhere learn to have in response to the dead and their remains.

Within British contexts, such expected social meanings and responses towards human remains are articulated in many ways, often contested. One apparently uncontested set of assumptions can be found in the very definition of human remains within the Human Tissue Act 2004 (section 53), which includes all material 'which consists of or includes human cells', but excludes hair and nail clippings. This definition is very obviously shaped by a particular social and historical moment: while the drafters of the Act and their advisers did not equate such materials with 'real' human remains and the need for care, many non-Western peoples still believe in the power of hair and nails to do magical harm, and are equally concerned with such materials as with other kinds of remains (Peers 2003). This set of meanings preserves the assumption that human remains are, or can be, agents within the social relationships amongst the living, whereas the scientific perspective articulated in the 2004 Human

Tissue Act has always attempted to deny or suppress such social meanings.

Indigenous Peoples and Human Remains

The HTA definition of human remains raises the question of what they mean for overseas indigenous people, especially to peoples who most often request to repatriate them: Native Americans, First Nations, Maori and Australian Aboriginal peoples. For such peoples, as for the parents involved in the Alder Hey and Bristol inquiries, human remains are kin to whom the living have a duty to mourn and to pay respect: to care for. In some cases, these are direct relatives: some indigenous groups have quite literally been looking for their great-grandparents' missing skulls. In other cases, as in the Northern Cheyenne who in 1993 retrieved the remains of Dull Knife's band – killed by US troops in 1879 – from the Smithsonian, they are lineal descendants (Thornton 2002: 16). In many more cases, tribal peoples claim cultural relationships with the dead: some *moko mokai*, tattooed and preserved heads of Maori warriors, cannot be specifically identified by clan or local group through provenance or tattoo markings, but have been claimed from British museums on the basis that they are simply Maori ancestors (eg. Kelbie 2004). In all of these cases, the living emphasize the social identity of the deceased as 'one of us', 'ours', 'our relative', 'our ancestor'.

While Hallam and Hockey (2001: 109) acknowledge that 'the negotiation of relations between the object body (the dead body in the present) and the embodied person (the living body of the past) is a complex process', they don't go far enough in considering indigenous remains. For many of the indigenous peoples considered here, there is little if any distinction between the 'object body' and the 'embodied person': those bones are a person. In North American societies of the northern Plains and Great Lakes, for instance, many groups believe that the spirit of the person stays with the remains, so that remains are often verbally addressed, using kinship terms, as if one is speaking to a living person; spirits may be given plates of food; the remains of children are often reburied with toys; and care is taken to identify the group or clan identity of remains so that members of the same group or clan can rebury them. For some tribes, the disturbance and removal of human remains causes damage both to the dead, whose journeys are interrupted, and to the living, whom they haunt (e.g. Riding-In 2000: 109); for other groups, the dead are potentially polluting and dangerous to the living, who have complex protocols to appease them: again, they are thought of as persons, dangerously powerful social agents, and are related through patterns of kinship and social relations. The nature of the continuing social identity and animacy of the dead is in part what drives repatriation claims, since from this perspective it is inappropriate and disrespectful to objectify remains through scientific and museological procedures including display and research, which place ancestral remains in what are often thought of as 'enemy' physical and intellectual contexts created by colonizers, and which are felt to strip social identity from them.

The meanings or realities of human remains for such peoples have much to do with the social status of the living who claim kin. Research in repatriation claims attempts to restore social identity to remains, to identify remains by region, clan or tribe, or even by name (as is also the case in forensic excavations following massacres, see Petrović-Šteger, in this volume). This reattachment of remains and social identity not only reasserts the group's own sense of identity and repudiates colonizers' categories of racial or ethnic identity or practices of anonymization, but allows the living to mourn for and engage in social relations with the dead. In such cases, the reassertion of kinship and social identity has another implication for the living as well: an assertion of their power to defy such problematic social and political relations and treatment, and to reinstate what they see as the proper patterns of social relations with the dead. Parents whose children's organs were retained without their consent at the Alder Hey and Bristol hospitals similarly fought for what they saw as their rights as parents to respect and mourn their children's bodies, against the power of the medical establishment and its refusal to acknowledge such social meanings of retained organs.[10] For tribal peoples in North America and the Pacific, claiming the bodies of the dead is part of broader attempts to lay the colonial past to rest, ending a phase when they were powerless to protect the bodies of their dead, and to reclaim autonomy in the present. Human remains in this context act as social agents in Gell's phrase (1998: 16), causing events to happen, sparking new forms of behaviour and relationships.

The Social Relations of Collecting Indigenous Remains

Most historic remains, of course, are anonymous; very few arrived in museum collections with names attached. This was a reflection of poor relations between source communities and collectors, so that the collectors were unable to establish the identities of the collected; it also reflected the importance of race or 'type' over that of individual identity for scholars. This aspect of the construction of scientific knowledge worked in tandem with the desire for anonymization of specimens for scientific research.

Those few remains which were of named persons reflected a colonial obsession with the salvage paradigm, being usually the bones of 'the last of his race', or her race, in the case of the Tasmanian woman Truganini (Fforde 2002: 28) – a hallmark of another problematic set of social and political relationships between collector and collected. The very removal of remains from burying grounds and often from the signs of social status which marked the graves of the dead further distanced them from their social identities (and see Hallam and Hockey 2001: 132). These removals from social contexts were quite deliberate acts, which translated physical remains from the local category of persons to the scientific category of specimens. Once in the museum, human remains were given new identities signalled by their classification within schema of racial types; their accession numbers, the information sometimes written on crania, and their cataloguing in museum records further distanced them from the persons they had once been. Like mourning rituals, which separate the living and the dead, acts of collecting, museum procedures and scientific analysis intervened between remains and persons.

This not-social, or even anti-social, perspective on indigenous remains was reinforced by the broader intellectual and political contexts in which they were collected. One of the crucial intellectual contexts was the widespread belief that 'authentic' indigenous cultures and peoples were dying out in the late nineteenth century. The collected were, in effect, already socially dead to their collectors, who were intellectually, politically and socially distanced from them. Death is feared in most societies because it causes social erasure (Hallam and Hockey 2001: 4), but the salvage paradigm in Victorian anthropology led to a neat double twist on this effect. As Hallam and Hockey note (2001: 44): 'rupture[s] in social interaction and communication can amount to social death prior to the event of biological death'. They are speaking about interactions with dying patients, but the same dynamics operated in colonial perspectives where there was little significant social interaction between tribal peoples and colonizers, where colonizers widely assumed that indigenous people were less fully human than they were and that, being inferior, they would soon die out. The operation of such beliefs constituted an important factor in the definition of indigenous remains as suitable for collection, and is still an important factor in debates over the fate of those remains today: no-one questions the return of remains of victims of Nazi medical experiments when these come to light,[11] but the remains of indigenous peoples have been deemed suitable to retain because of their relative lack of power and social presence in Britain. The difference is not in the nature of the human remains involved, but in the social relations in which they have existed and continue to exist, and in which their own scientific interest, evoking a putative status of asocial remains, can persist.

Such 'ruptures in social interaction' leading to social death and powerlessness were certainly apparent in the circumstances of collection of indigenous remains by British colonial collectors – circumstances which would never have been pursued with groups regarded as social equals. Many specimens were obtained by theft from graves and tombs, crude excavations at night, or bribes (see, for example, Cole 1995[1985]:171, 175, 307–8; Fforde 2002: 26–28). Anthropologist Franz Boas, who collected hundreds of sets of skeletal remains on the Northwest Coast of Canada, wrote that stealing bones from graves was 'repulsive work', but also that 'someone had to do it' (cited in Cole 1995[1985]: 308). Some of his collections were traded to British museums. In Australia, the point is made bluntly: 'The building of such collections involved grave robbing, contract killing, massacres and murder. It has been estimated that the graves of 5,000 to 10,000 Aboriginal Australians were opened, the bodies dismembered and parts stolen for scientific studies ... Aboriginal deaths were often the result of massacres connected with the dispersal of indigenous settlements ...' (Jones and Harris 1998: 261). To indigenous peoples, ancestral remains held in overseas museums reference such appalling social relations: the bones, preserved soft tissue, hair, nails, and other remains' collected in coercive circumstances and stripped of social identity in their scientific contexts are a reminder of what their people went through, a site of social memory, and a history that needs to be laid to rest just as much as the ancestors themselves.

British Meanings of Indigenous Human Remains

For British collectors, indigenous human remains have been part of a much larger group of signifiers of human difference and as equally 'enmeshed in the social relations of power' (Lindman and Tarter 2001: 2) as bodies in any other state. The remains of indigenous peoples have long-established meanings deriving from a history of interactions between England and its colonized peoples. Equally, they have meanings deriving from ancient beliefs about physical and cultural 'others' who have been understood as mirrors of ourselves: the variously positive and negative constructions of the Savage and the Noble Savage, as the antithesis of civilized humanity (e.g. Dickason 1984). The bodies and body parts of these 'wild', 'savage', and colonized peoples have similarly been regarded as Other, and have played an important role in the construction of systems of knowledge which have posited hierarchical relationships between these peoples and British Anglo-Saxons (on which, see Chapman 1978).

Within emerging biological and social science, collections of human remains became central in the early modern period to the development of theories about the origins of human populations and the relationships between them. In particular, crania were collected from populations around the world to facilitate comparison of physical and intellectual capacity of different peoples, and to study the relations between groups (see, e.g. Stocking 1987, 1988; Zimmerman 1997: 95; Fforde 2002: 25–26). Such collections were expected to include indigenous specimens.

Within Britain, collections of human remains from around the world functioned in a related way, as signifiers of difference between British people and those they sought to colonize. Widespread discourses within eighteenth-and-nineteenth century British society used many forms of evidence about the nature of Others, including live captives brought back by explorers; moralistic plays and ballads about such individuals and their societies (e.g., a popular play about Omai, who accompanied Cook back to England, Wilson 2003: 63); portraits of such individuals; physical measurements; skeletal remains and measurements taken from them; anthropometric photographs; measurements of hair colour and texture; measurements of skin colour; and displays of living indigenous peoples at World's Fairs. As Wilson (2003: 63) has noted, such forms of evidence tended to essentialize identities and create and articulate idealized, hierarchical relationships between British people and 'savages'. Human remains, or images and measurements, were often displayed in conjunction with ethnographic artefacts from the same culture or region, to further emphasize the point that those displayed were 'primitive' culturally as well as physically. Physical traits were also linked to intellectual and moral qualities and to assumptions about hierarchical relationships between peoples. Scientific techniques of analysis, and the dissemination of information through scholarly and popular venues, authenticated narratives of race and legitimated popular beliefs within Britain in the superiority of British people over others (Coombes 1994; Harrison 1995: 52).

The stories told about bodies were crucial in the process of England explaining the world to itself, categorizing other peoples vis-à-vis themselves. The use of human remains from Tasmania, Australia, Africa and the Americas in defining 'race' was an important part of this process. Intriguingly, these collections of remains of people defined as quintessential 'Others' came to act as a sort of mirror, to define what was as well as what was not British. Indeed, historian Linda Colley has argued (1992: 6) that the British in the eighteenth and nineteenth centuries 'came to define themselves as a single people not because of any political or cultural consensus at home, but rather in reaction to the Other beyond their shores'. I would modify this to suggest that the development of modern British identity – a sense of 'us' – occurred within Britain as well, in relation to the encounter, classification and representation of many 'thems'. Collections of indigenous human remains played a significant part in this process.

Karp (1992: 1) has argued that 'the selection of knowledge and the presentation of ideas and images are enacted within a power system. The sources of power are derived from the capacity of cultural institutions to classify and define peoples and societies.' Making some material and ideas 'socially visible' (Hallam and Hockey 2001: 9) to certain groups – displaying the idea of racial hierarchies to precisely the middle and upper classes of British society who were participating in colonial regimes – while suppressing other material and ideas, in the particular context of the British Empire, meant that collections of indigenous human remains in Britain were never just specimens, but were wrapped in layers of social and political meaning. Collections of the remains of colonized people were used to formulate ideas about their inferiority to British colonizers and thus legitimate the imposition of colonial control (Stocking 1991, 1988: 7; Fforde 2002: 29). In such contexts, theories about race served not only to reinforce boundaries of identity, but served to 'establish a hierarchy of social relations' (Lindman and Tarter 2001: 4). The social relations established around indigenous human remains during the colonial era and in colonial contexts are only beginning to shift now, and the remains themselves are proving to be a significant catalyst for this change.

Fighting over the Body (Parts): Debates in Britain

A new set of attitudes, practices and relations is emerging around indigenous remains between British scientists, museum professionals and source communities. This emerging praxis concedes that source communities have rights in the care, treatment and retention of ancestral remains, and that requests for the return of remains to descendants should be honoured. This position is contested fiercely by conservative elements in Britain who decry the loss of potential scientific data, right of research, and control over collections that the new praxis advocates: the meanings of these remains are especially unstable in this first decade of the twenty-first century. The two positions emerged strongly in the final report of the WGHR, which largely advocated the former, with a dissenting opinion representing the latter views.

The UK debates about indigenous remains have at all levels – scholarly, governmental and in the popular media – been characterized by an almost complete absence of indigenous peoples and by the strong presence of scientific voices and institutions. Despite the centrality of indigenous groups to the matter at hand, DCMS did not provide funds to consult systematically with them during the Working Group's submission period.

Only four indigenous groups or individuals, and one 'bicultural' museum, gave evidence to the group (out of forty-seven submissions in total). Nor did the media feel the need to include the words or views of indigenous peoples when reporting on the group's activity. Of several dozen national media articles on the report of the Working Group on Human Remains, only four included short quotations from indigenous peoples. All articles included strong voices from staff at the two largest collections of indigenous remains in England, the Leverhulme Centre for Human Evolutionary Studies in Cambridge and the Natural History Museum in London. Scientists quoted insisted that the potential research benefits of collections of human remains were being downplayed by the Working Group, and several made statements implying that indigenous claimants had no legitimate social or biological basis on which to claim rights in collections – thus positing scientific research as the most rightful 'owner'.[12]

This absence of indigenous people from the debates is one manifestation of a vast social and political distance between British museum professionals and scientists on the one hand, and indigenous peoples on the other, which persists from colonial contexts. Borrowing from Johannes Fabian (1983), such distance constitutes a denial that peoples represented by indigenous human remains exist in the same temporal space as scholars themselves. The space between them is a conceptual one, a denial of social co-existence, of equality. As Fabian pointed out, such denial is a maneuver to create hierarchical relations of power over the people studied: in this case, a way of constructing a space in which the scholar controls not only the collections of remains, but the knowledge derived from them. In this model, the source communities have no rights over collections of human remains.

There are parallels here with the perceived paternalistic relationship between medical science and patients which was fundamentally challenged by the inquiries following the Alder Hey and Bristol retention scandals, when the right of patients to consent to collection and retention of specimens surfaced. In regards to indigenous human remains, the claims by scientists and museums that they have the right to continue to hold remains without community consent carries over from colonial relationships, in which indigenous people were in effect socially and politically dead (absent, powerless, voiceless) to colonial collectors and scholars. Refusing to engage with indigenous communities over the disposition of human remains signals a desire within the museum profession and scientific community to maintain long-standing identities and group boundaries, including professional ones, as well as control and authority over collections of remains. The force and passion of the debates within Britain over the disposition of indigenous human remains is one articulation or performance of core beliefs underpinning identity and relative hierarchies in a post-colonial context.

The distance maintained by British holding institutions and government from indigenous groups in these developments is very different from that in North America and the Pacific, where institutions with collections of indigenous remains have actively sought to develop new social relations with source communities to provide guidance for the care, research and possible return of remains. The existence of such continuing poor relations is articulated in the most commonly voiced phrase heard in the UK in response to the idea of working with source communities: 'But won't *they* want it all back?' This question speaks volumes about the social and political distance between 'us' and 'them', about a profound distrust of 'them' (who were defined as such partly through the historical analysis of human remains) and of a widespread fear of loss: not just the loss of collections, but of colonial power and authority over collections. In this sense, indigenous remains function for some British as material touchstones which are 'instrumental in the maintenance of particular memory configurations' (Hallam and Hockey 2001: 7): a memory of power, and of colonial relations with the collected.

The potential loss of such touchstones is especially threatening to certain professions (museological, scientific) which rely on collections assembled through such relations. The suggestions in the report of the WGHR (e.g. p. 112, 170) that holding institutions should obtain consent from source communities as a basis for continuing to retain collections of indigenous humans was widely denounced by these groups (Chalmers 2003: 182–83; Fazackerley 2003). I am reminded of Verdery's observation (1999: 33) that 'dead bodies ... can be a site of political profit'. For such stakeholders, the idea of negotiating new social relations with indigenous peoples around collections of human remains is profoundly threatening.

The relative absence of indigenous peoples from the British debates similarly underscores the threat that the sociality of bodies poses to 'scientific collections'. Social identity, in the form either of a relationship of equality with an indigenous community, or as a specific name attached to human remains (as in some crania in several museum collections which have the names of the deceased persons from whom they came written on the front of the crania) translates human remains from scientific data to ancestors, and potentially negates their use for scientific research. Such evidence is always present – meanings of human remains are always co-present – but usually, in the context of the laboratory or storeroom, scientists and museum professionals suppress the social implications of such evidence in favour of their own sets of meanings. Consultation with indigenous peoples reminds us that such meanings might be suppressed, but they are still present (and see Peers 2003 on such a consultation).

Having a name attached to the remains makes it virtually impossible (for legal and moral reasons) to refuse repatriation requests. With names, human remains become social persons, with demonstrable kin who are legitimate claimants in law.[13] Names are thus key not only to re-establishing the social identity and meanings of remains, but to renegotiating relationships between museums and source communities, destroying the social vacuum in which they have been retained and studied: these specimens become people, with kin and descendants whom one cannot turn away.[14]

Some British institutions (the Royal College of Physicians and Surgeons; the Royal Albert Memorial Museum, Exeter; Manchester Museum; Glasgow Museums; and others) have been quite willing to treat indigenous human remains as social, and to deal with the implications of this. Some of these have directly articulated a felt need to alter colonial relations through the act of repatriating human remains.[15] Many other museums have grasped the idea that it might be beneficial to work with source communities on a range of projects, including the development of policy for handling, photographing, doing research on, displaying and storing human remains. This has sometimes led to controversial decisions to restrict access to collections of human remains, to store them separately from other materials, to insist on certain standards for storage.[16]

Indigenous remains, coming from colonized and still relatively powerless groups, within state-sanctioned museum and scientific institutions of a former colonial power, arenas for the validation (or contestation) of authority and legitimacy within our social world, thus become catalysts for uneasy debate in Britain over colonial history and its legacies: this is presumably one of the reasons for the relative absence of indigenous people within these debates, which constitute a process of working through internal questions within Britain. Intriguingly, the question of whether to treat these remains as persons or as specimens, as social or not, raises questions about the nature of British society and its relationships past and present with formerly colonized peoples. The possible repatriation of human remains constitutes a serious challenge to the politics of knowledge and ownership, for as James Clifford (2004: 18) has stated, it 'establishes indigenous control … and thus the possibility of engaging with scientific research on something like equal terms'.

Renegotiating Relationships – or Not

If this volume is interested in 'what happens to sociality in the face of new types of transaction that focus squarely on elements of the corporeal'

(Introduction, p. 2), then repatriation becomes interesting as a type of transaction which involves body parts, but is profoundly social in nature.

Repatriation is not just a way of acknowledging problematic social and political relations in the past. The return of human remains to source communities is a way of creating new relationships and, through this in turn, of changing the way museums and scientists think and work. Repatriation might be thought of as a form of gift exchange in the classic anthropological sense in which the movement of valued objects from one group to another (whether it is between clans in marriage exchange, or between nations as in diplomatic gifts) actually creates and maintains relations between peoples. The act of the gift embodies the values of hospitality, generosity and the reciprocal relations of kinship which are desired in the relationship, and helps to define appropriate behaviour by parties within these relationships afterwards. The gift is not just a single act of exchange, but (in Mauss's phrase) a total social fact, an encompassing process. As Humphrey and Hugh-Jones, following Strathern, note regarding exchange: 'What is exchanged are not things for things, or the relative values of people quantified in things, but mutual estimation and regards'.[17] The gift of human remains from holding institutions to indigenous claimants is (ideally) a demonstration of respect for the living, an acceptance of the legitimacy and authority of the claimant. Such events are often the first time since the time of collection that people have spoken to each other cross-culturally about the meanings of human remains, and have tried to understand and respect each other's perspectives.

Repatriating human remains has always had to do with relations amongst the living. In European contexts, the repatriation of the dead as well as of prisoners of war has been a crucial element in restoring positive relations among nations in the aftermath of war, a situation perhaps not incomparable to that regarding indigenous remains (Barkan 2002: 18–19). Verdery's observation (1999: 19) that, 'By repositioning them [i.e. bodies of leaders/ancestors], restoring them to honour, expelling them, or simply drawing attention to them, their exit from one grave and descent into another mark a change in social visibilities and values.' Verdery is writing of postsocialist transformations; the repatriation of indigenous remains has been about postcolonial ones.

With the amendment to the British Museum Act forced by the HTA, allowing the national museums to release human remains,[18] the way has been paved for the creation of new social relations as remains are passed back to source communities. Other holding institutions in Britain did not face this same legal barrier to repatriation. While many museums and collections have already repatriated, though, it is difficult to say whether new relations have actually been forged, or whether these have simply been symbolic, only briefly social, acts in which remains have been transferred to the control of claimants. While virtually all handover ceremonies for human remains involve statements about the moment heralding a new era for such relationships, I know of no case at all in which a former claimant community continues to be involved in any way with an institution which has handed remains back. Indeed, internal discussions with colleagues at some British museums suggest that claims for the return of human remains are regarded as potentially damaging for the museum's public image, and so the institution is glad to return the remains – and to not have further contact with claimant groups. While repatriation can potentially create new forms of sociality around human remains, then, it hasn't yet begun to do so significantly in British contexts.

Conclusions

In thinking about the effects of material culture, Chris Gosden has claimed that 'our sense of self as individuals and as groups is built up through objects which connect us to others, or isolate us as individuals' (2003: 181). In British contexts, human remains have functioned very much to create a sense of a British self. I want now to consider other work that such collections do socially.

Although the acquisition and analysis of indigenous human remains has advanced scientific knowledge, these processes have also created structures of knowledge we now understand as racist. At one level, such collections and their use might be understood, in Gananath Obeysekere's phrase, as 'enactments of racial intolerance', creating and maintaining social boundaries between peoples.[19] I would argue that the continued retention and intellectual use of indigenous human remains in Britain, in the absence of and refusal (or failure) to create relationships with source communities, can be seen this way. One might also see the refusal to acknowledge other cultural perspectives on human remains, the refusal to consult with overseas communities about them, and the insistence by the scientific community on retaining control over such remains, as an enactment of social intolerance, an antisocial act, equivalent in many ways to the forms of magical harm which tribal source communities have always assumed could be wrought using hair and nails.

Following this train of thought, I have sometimes wondered about the 'Treatment of Dead Enemies' case in the Pitt Rivers Museum, and about the layers of meaning and relationships embodied by the display. As I noted in the new text I prepared for the case, the taking of enemy heads and scalps has, at certain moments in history, been sanctioned by many societies worldwide. These forms of violence, I wrote, effect and maintain social order. I wonder if this works in a double sense, however: if the heads and scalps in this case do not merely exemplify what the Shuar, or Naga, or Blackfoot, did to their enemies, but still represent to us and our visitors the racialist hierarchical order in which they were deemed to have meaning when they were collected. For many indigenous peoples, such theories, and the museum spaces in which they were displayed, were enemy spaces, intellectually and politically. And then I wonder if continuing to display their remains in this case gives its label another meaning. This is not just about what other peoples did to dead enemies, but also about what we in contemporary Britain think of these peoples whose remains are displayed within it, and how we position ourselves in relation to them: about maintaining a distanced and colonial gaze. The material in the 'Treatment of Dead Enemies' case is both constituted and contained not only by wood and glass but by historical social relations over time. The current debates about the disposition of human remains could be seen as having much to do with challenging – or maintaining – a long-established social and political order, the boundary-reinforcing functions that human remains have long been required to effect.

Gell (1998) and Gosden (2003) have also argued, however, that artefacts also function not just as boundaries, but as nexuses of relationships, as points of contact, interaction and exchange around which relationships are constructed, maintained and renegotiated. Collections of human remains function in this way across their 'lives', which extend long past death. For those of us in Britain who do have social relationships with indigenous peoples, human remains activate those relationships. At the request and guidance of many tribal members in North America, when I work with human remains from their communities, I speak to the spirits associated with them: I offer tobacco (a common ritual offering in many North American tribes) to these powerful spirit beings, and I address them in respectful grandparent terms, apologizing for disturbing them and saying that I hope that what I do will benefit them and their people, and telling them that they are not forgotten. I often feel awkward when I do so, standing in museum spaces far from tribal communities, but I recall specific conversations with tribal people about the need for such acts. What I have chosen to do honours my relationships with living tribal members, as well as honouring their sense of social relations with these dead.

These remains act as powerful social agents creating new relations between peoples today. As Lambert and McDonald state in the introduction to this volume, '… acts of bodily fragmentation, dismemberment, transfer, reassignment or transformation are never confined solely to the [biological] … in their effects but inevitably entail the reformulation, reconstruction or re-establishment of social relations between persons and groups'.

While the transfer to museums and scientific institutions enmeshed indigenous remains within scientific discourses which sought to strip their earlier social meanings from them, such emerging praxis reminds us that this process of desocializing has been unsuccessful. The fierce debates within Britain recently and numerous repatriation cases of human remains have leaned significantly towards a definition of historic indigenous remains as social, with social meanings and relationships in which the very different practices and relationships of scientific collecting and display had intervened, and which they had muted. Whether repatriation will come to constitute a new form of sociality, however, or what other forms will emerge around these remains, is yet to be seen.

Acknowledgements

Many thanks for stimulating ideas to members of the WGHR, and to: Marcus Banks, Maurice Davies, Elizabeth Edwards, Chris Gosden, Tamasin Graham, Cara Krmpotich, Frances Larson, Maureen Matthews, Mike O'Hanlon, Laura Rival, Peter Riviere, Natalia Shunmugan, Elizabeth Vibert and David Wengrow. Helen Lambert and Maryon McDonald offered encouragement and helpful editorial suggestions.

Notes

1. I would like to thank Michael O'Hanlon, Peter Rivière, Laura Rival, Anne Christine Taylor, Peter Gordon and Ollie Douglas who assisted me with this project.
2. The Retained Organs Commission report is at: www.nhs.uk/retainedorgans/. The Church Archaeology Human Remains Working Group report is at: www.englishheritage.org.uk/default.asp?wci=Node&wce=8246. The report of the Working Group on Human Remains (WGHR) is at: www.culture.gov.uk/global/publications/archive_2003
3. The shrunken heads are the subject of periodic media attention, the stories usually focusing on perceived threats to their continued display; see, for instance, 2007 media coverage and strong reader responses to it: http://www.theoxfordtimes.net/mostpopular.var.1191233.mostviewed.should_shrunken_heads_stay_in_museum.php
4. The Museum Ethnographers' Group Guidelines on Management of Human Remains, 1991, is reprinted in WGHR Report (2003: 263–65). The Human Tissue Act 2004 (http://www.opsi.gov.uk/acts/acts2004/20040030.htm) has strict guidelines for the display of human remains. The Department of Culture, Media and Sport has also issued a Guidance for the Care of Human Remains in Museums (2005).

5. I use the definition of 'indigenous' as adopted by the UN Working Group on Indigenous Population: a priority in time, with respect to the occupation of a specific territory; the voluntary perpetuation of cultural distinctiveness; self-identification, as well as recognition by other groups as a distinct collectivity; and an experience of subjugation, marginalization or dispossession (cited in Simpson 1997: 22).
6. The North American debates are summarized in Bray and Killion (1994), Mihesuah (2000), Fforde et al. (2002).
7. I use the term 'source communities' to designate the peoples from whom human remains and ethnographic artefacts were collected. See Peers and Brown (2003).
8. See the Report of the Royal Liverpool Children's Inquiry, 2001, at: http://www.rlcinquiry.org.uk/; the Report of the Bristol Royal Infirmary Inquiry, 2001, at: http://www.bristol-inquiry.org.uk/; and the Isaacs Report (Retained Organ Commission), 2003 and related government guidance on postmortem practice at: http://www.dh.gov.uk/PublicationsAndStatistics/
9. I borrow the phrase 'ontologically unstable' from Elizabeth Edwards (2001: 5), who (citing work by Christopher Pinney) uses it regarding anthropological photographs. Like photographs, which can have multiple sets of meanings attached to them depending on the viewer and the context, human remains refuse to settle in one box of meanings.
10. See, on this, Hall (2001).
11. The remains of children who had died under the Nazi regime were discovered in an Austrian hospital in 2002. The remains were immediately identified and returned to families (Miller 2002).
12. Chris Stringer of the Natural History Museum, in an article in the *Telegraph* (12 November 2003), claimed that, 'We could see whole fields closed off to research if we lose key specimens' and insisted that any approach to human remains must emphasize their scientific importance 'and potential benefits'. Similarly, Sir Neil Chalmers, then head of the Natural History Museum, insisted throughout the Working Group's process that 'You have to balance the benefits of research against [the rights] of claimant groups' (quoted in Morris 2003).
13. On the issue of names and human remains, see also Hubert and Fforde (2002: 12). While otherwise strongly retentionist, the Duckworth Laboratory in Cambridge has made it policy 'to return any individual skeletons or skulls of named individuals' (WGHR 2003: 60).
14. Hallam and Hockey (2001: 89–90) give a British parallel for issues of names and sociality and politics surrounding bodies: 'While communities of commemoration construct highly differentiated social identities for the dead, the threat that the dead body will be reduced to an anonymous materiality remains a disturbing prospect. This is evidenced in public controversies regarding the treatment and places of corpses in medical institutions. When a photograph of bodies temporarily stored on the floor of a hospital chapel (Bedford, England) was published in national newspapers in January 2001 … it signalled and provoked anxieties about the assimilation of persons into a

collective mass of bodies – in this hospital space they were now "objects" denied their subjectivity.'

15. See, for instance, statements by the heads of the Royal College of Surgeons and the University Museum of Manchester on the handover of human remains from their institutions, cited in WGHR Report (2003: 53–56).

16. Tiffany Jenkins has been the most vocal opponent of such decisions; see Jenkins (2002, 2005).

17. Humphrey and Hugh-Jones (1992: 17), discussing a paper in the same volume by Strathern.

18. Section 47 of the Human Tissue Act 2004 gives national museums in Britain the power to de-accession human remains.

19. Obeyesekere, keynote address, Association of Social Anthropology meeting 2003, Manchester.

References

Barkan, Elazar. 2002. 'Amending Historical Injustices: the Restitution of Cultural Property – an Overview', in Elazar Barkan and Ronald Bush (eds), *Claiming the Stones, Naming the Bones: Cultural Property and the Negotiation of National and Ethnic Identity*. Los Angeles: Getty Research Institute.

Bray, T.L. and T.W. Killion (eds) 1994. *Reckoning with the Dead: the Larsen Bay Repatriation and the Smithsonian Institution*. Washington, DC: Smithsonian Institution Press.

Chalmers, Sir Neil. 2003. 'Statement of Dissent.' *Report of the DCMS Working Group on Human Remains*.

Chapman, M. 1978. *The Gaelic Vision in Scottish Culture*. London: Croom Helm.

Clifford, James. 2004. 'Looking Several Ways: Anthropology and Native Heritage in Alaska', *Current Anthropology* 45(1): 5–30.

Cole, Douglas. [1985] 1995. *Captured Heritage*. Vancouver: UBC Press.

Colley, Linda. 1992. *Britons: Forging the Nation, 1707–1837*. New Haven, CT: Yale University Press.

Coombes, Annie E. 1994. *Reinventing Africa: Museums, Material Culture and Popular Imagination in Late Victorian and Edwardian England*. London and New Haven, CT: Yale University Press.

Cove, John J. 1995. *What the Bones Say: Tasmanian Aborigines, Science and Domination*. Ottawa: Carleton.

Dickason, Olive. 1984. *The Myth of the Savage*. Edmonton: University of Alberta Press.

Edwards, Elizabeth. 2001. *Raw Histories: Photographs, Anthropology and Museums*. Oxford: Berg.

Fabian, Johannes. 1983. *Time and the Other: How Anthropology Makes its Object*. New York: Columbia University Press.

Fazackerley, Anna. 2003. 'Human Remains Code is Unworkable', *Times Higher Education Supplement*, 21 November 2003, p. 3.

Fforde, Cressida. 2002. 'Collection, Repatriation, and Identity', in Cressida Fforde, Jane Hubert and Paul Turnbull (eds), *The Dead and their Possessions: Repatriation in Principle, Policy and Practice.* London: Routledge, pp. 25–46.

Gell, Alfred. 1998. *Art and Agency.* Oxford: Oxford University Press.

Gosden, Chris. 2003. 'Object Lessons and Wellcome's Archaeology', in Ken Arnold and Danielle Olsen (eds), *Medicine Man: the Forgotten Museum of Henry Wellcome.* London: British Museum Press.

Hall, David. 2001. 'Reflecting on Redfern: What Can We Learn from the Alder Hey Story?', *Archives of Disease in Childhood* 84: 455–56(June). Available online at: http://adc.bmjjournals.com/cgi/content/full/84/6/455

Hallam, Elizabeth and Jenny Hockey. 2001. *Death, Memory and Material Culture.* Oxford: Berg.

Harrison, Faye. 1995. 'The Persistent Power of "Race" in the Cultural and Political Economy of Racism', *Annual Review of Anthropology* 24: 47–74.

Hubert, Jane, and Cressida Fforde. 2002. 'Introduction: the reburial issue in the twenty-first century', in Cressida Fforde, Jane Hubert and Paul Turnbull (eds), *The Dead and their Possessions: Repatriation in Principle, Policy and Practice.* London: Routledge.

Humphrey, Caroline and Stephen Hugh-Jones. 1992. 'Barter, Exchange and Value', Introduction to Humphrey and Hugh-Jones (eds), *Barter, Exchange and Value: An Anthropological Approach.* Cambridge: Cambridge University Press.

Jenkins, Tiffany. 2002. 'Sense and Sensitivity', *The Spectator*, 7 September 2002.

———— 2005. 'The Censoring of Our Museums', *New Statesman*, 11 July 2005. Available online at: http://www.newstatesman.com/Arts/200507110035

Jones, D. Gareth and Robyn J. Harris. 1998. 'Archaeological Human Remains: Scientific, Cultural and Ethical Considerations', *Current Archaeology* 39(2): 253–64.

Karp, Ivan. 1992. 'Introduction: Museums and Communities: the Politics of Public Culture', in Ivan Karp, Christine Mullen Kreamer and Steven D. Lavine (eds), *Museums and Communities: the Politics of Public Culture.* Washington, D.C.: Smithsonian Institution, pp. 1–17.

Kelbie, Paul. 2004. 'Maori Win Return of Preserved Heads from Glasgow Museum', *New Zealand Herald*, 26 June 2004. Online at: http://www.nzherald. co.nz/topic/story.cfm?c_id=350&objectid=3574540

Lindman, Janet Moore and Michele Tarter (eds). 2001. *A Centre of Wonders: the Body in Early America.* Ithaca, NJ: Cornell University Press.

Lyons, Claire. 2002. 'Objects and Identities: Claiming and Reclaiming the Past', in Elazar Barkan and Ronald bush (eds), *Claiming the Stones, Naming the Bones: Cultural Property and the Negotiation of National and Ethnic Identity.* Los Angeles: Getty Research Institute, pp. 116–37.

McGuire, Randall H. 1992. 'Archaeology and the First Americans', *American Anthropologist* 94(4): 816–36.

Mihesuah, D.A. (ed.). 2000. *Repatriation Reader: Who Owns American Indian Remains?* Lincoln: University of Nebraska Press.

Miller, Barbara. 2002. 'Child Remains from Nazi Euthanasia Clinic Laid to Rest', *Independent*, 25 April 2002, p. 11. See also: http://news.bbc.co.uk/2/hi/

europe/1956494.stm; http://www.time.com/time/europe/magazine/2000/
0403/trial.html; http://www.news.com.au/common/story_page/0,4057,
4214012%5E401,00.html

Morris, Jane. 2003. 'Dead but Not Buried', *Museums Journal* December 2003.

Peers, Laura. 2003. 'Strands Which Refuse to be Braided: Hair Samples from Beatrice Blackwood's Collection at the Pitt Rivers Museum', *Journal of Material Culture* 8(1): 75–96.

Peers, Laura and Alison K. Brown (eds). 2003. *Museums and Source Communities: A Routledge Reader*. London: Routledge.

Richardson, Ruth. 2001. *Death, Dissection and the Destitute*, 2nd edition. Chicago: University of Chicago Press.

Riding-In, James. 2000. 'Repatriation: a Pawnee's Perspective', in Devon Mihesuah (ed.), *Repatriation Reader: Who Owns American Indian Remains?* Lincoln: University of Nebraska Press.

Simpson, Tony. 1997. *Indigenous Heritage and Self-Determination*. Denmark: International Working Group for Indigenous Affairs (Forest Peoples Programme).

Stocking, George W. Jr. 1987. *Victorian Anthropology*. NY: Free Press.

——— (ed.). 1988. *Bones, Bodies, Behavior: Essays on Biological Anthropology. History of Anthropology*, vol. 5. Madison: University of Wisconsin Press.

——— 1991. *Colonial Situations: Essays on the Contextualization of Ethnographic Knowledge*. Madison: University of Wisconsin Press.

Stringer, Chris. 2003. 'Bones of Contention', *Telegraph*, 12 November.

Thornton, Russell. 2002. 'Repatriation as Healing the Wounds of the Trauma of History: Cases of Native Americans in the United States of America', in Cressida Fforde, Jane Hubert and Paul Turnbull (eds), *The Dead and Their Possessions: Repatriation in Principle, Policy and Practice*. London: Routledge, pp. 17–24.

Verdery, Katherine. 1999. *The Political Lives of Dead Bodies: Reburial and Post-socialist change*. New York: Columbia University Press.

Wilson, Kathleen. 2003. *The Island Race: Englishness, Empire and Gender in the Eighteenth Century*. London: Routledge.

Working Group on Human Remains (WGHR). 2003 Report. Department of Culture, Media and Sport. Available at: www.culture.gov.uk/cultural_property/wg_human_remains/default.htm

Zimmerman, Larry. 1997. 'Anthropology and Responses to the Reburial Issue', in Thomas Biolsi and Larry Zimmerman (eds), *Indians and Anthropologists: Vine Deloria Jr. and the Critique of Anthropology*. Tucson: University of Arizona Press.

Chapter 4

TOWARDS A CRITICAL ÖTZIOGRAPHY: INVENTING PREHISTORIC BODIES

John Robb

Introduction: Resocializing Unknown Bodies

This paper began its life as an attempt to answer a naïve question, 'Why does the Ice Man have to have a name?' I quickly became aware, however, that this apparently simple question takes us very quickly into deep waters: it appears simple only because, as natives of our own culture, we take all the complicated issues underlying it for granted. As Helen Lambert and Maryon McDonald point out (Introduction, this volume), the body is always recognized and understood within social relations. A human body is never a neutral, purely material object. Given this, finding an unknown body, without any social ties or context, is like being presented with a Rorschach blot: it forces one to re-enact freshly one's foundations. The unexpected appearance of a body, with no social antecedents or biography, compels those present to match its appearance with the social invention of the same body. A chain of events which is triggered by such an appearance of a body thus presents us with a remarkable opportunity for observing our own or other people's reflexes where bodies in general are concerned.

Enter, thus, the Ice Man. 'Ötzi', the 'Ice Man', is the naturally mummified body of a man who died high in an Alpine pass on the border of modern Italy and Austria sometime between 3300 and 3000 BC. One of the most spectacular archaeological discoveries of recent times, the Ice Man has been known to us for fifteen years now. In that time, his naked and battered body has not only been probed and scrutinized with virtually every invasive and non-invasive technique of which science can dream, but he has been the subject of an incredible range of speculation and innuendo. Scarcely a year goes by without some new revelation or palaeo-scandal. He has been cast in the role of a bewildered frozen and inept mountaineer, a murderer, a murder victim, a ritual sacrifice, a fake. There has been speculation that he was an Egyptian, a eunuch, a practising homosexual and, recently, a genetically infertile social outcast.

One might dismiss all this as a frothy excess of bad taste, regrettable but only to be expected of the popular press. But my thesis here is that Ötzi deserves anthropological consideration – not only as a uniquely preserved fourth-millennium body which can tell us much about prehistoric Europe, but also as a construction of modern imagination which can tell us much about how contemporary, but usually implicit, understandings of the human body permeate visions of the past. The goal of this paper, thus, is a critical Ötziography, a study of how archaeologists have written Ötzi, the Ice Man of the Alps. How we have understood the Ice Man's body reveals much about how we understand our own bodies.

The Agency of Dead Bodies and the Need to Repersonalize

Dead bodies have power; sometimes far more than they had when they lived. They exert agency (Williams 2004). Exemplary corpses, dismembered, drawn, quartered, beheaded, burnt and displayed, enforce the power and legitimacy of the state. The dead protest: as with Antony's oration over Caesar's body, or the blood-clotted clothes of Hoskuld in *Njal's Saga*, the relics of victims are sometimes used to incite vengeance. Funerals of IRA members in Northern Ireland were flashpoints for rebellion; more recently, the photograph of a dead victim of Hurricane Katrina lying abandoned in the streets of New Orleans was more powerful in compelling the American government to act than anything that could be said on the Senate floor. Dead figures of state gather the good and the great to see them off. Nor is this ability to initiate chains of event found only with the eminent or unusual dead. Even in everyday death, the corpse cannot stay where it is, but must be disposed of, and it must be disposed of properly and well. Whole specialized industries exist to deal with dead bodies, whether lying quietly in a hospital bed or turning up unexpectedly in a field or cellar, and recourse to these institutions is not optional. Response to corpses is a mandatory and highly structured act. It is essential to realize that these are not merely cases of the living using dead bodies for pre-existing purposes such as furthering a political ambition which might otherwise be accomplished via some other gesture. Rather, it is the very presence of the dead body and the way in which people understand it which creates the situation to which people respond and the range of appropriate responses.

But can dead bodies (or for that matter the disembodied dead – ghosts and ancestors) really have agency? The answer has to be yes. In ordinary speech, agency is often equated with a capacity for autonomous, volitional action, characteristics which, in Western ontology, are restricted to beings which are understood to be biologically living. But agency is not a unilateral, essential quality which can be possessed, contained or used. Rather, it is relational; it exists only as it is defined within a particular social relation, whether between people or between people and things (Gell 1998). Hence the paradox that material things exert agency, in the sense of shaping the course of unfolding events, but they do so only by virtue of how people understand them. As Peers (this volume) notes, 'human remains themselves are so profoundly wrapped in social meanings that they not only act as extensions of the people they were, but as powerful social agents engaged in ongoing social relations'. The dead, thus, exert an agency with which they have been endowed by the living.

The Corpse and Its Agency

A fundamental component of the European tradition, often ascribed variously to Plato, Christian theologians and Descartes, is the long-standing Western and Christian opposition of the body and the soul. In contrast to many other traditions (cf. chapters in this volume by Peers and by Vilaça), European theologians and philosophers have traditionally understood the person as made of two distinct components, one spiritual, rational or intellectual and the other earthly and material.[1] The human body thus consists of a bounded material entity occupied or, literally, animated by a reasoning volition (Grosz 1994). Building upon this, in modern Western Europe, and particularly since the Enlightenment, persons have been understood as equivalent to the bounded physical body (Van Wolputte 2004). While this view has been grounded in anthropological comparison (Strathern 1988, 1996), it is also plausible in light of a wide range of bodily technologies from within our own society, for example the importance of body form and consumption as creators of self-identity (Shilling 2003). Similarly, bioethics and medical anthropology problematize such things as organ transplantation, cloning, stem cell research, gene therapy, and so on. What these new technologies have in common is that they transgress the boundaries of individual bodies or threaten their uniqueness or autonomy, something we find problematic (Kaufman and Morgan 2005). For example, Kaufman et al. (this volume) show how corporeal transactions between bodies via organ donations are experienced and expressed as social relatedness and affection.

This equation of the social person with the bounded, autonomous individual body permeates a modern relationship to dead bodies. As Tarlow (1999) has argued, in early modern Britain, when the bipartite person died, it was thought that the spirit and body separated; the soul was liberated from earthly shackles to its transcendental destiny. Consequently, the decay of the body was a normal part of the process; dead bodies in themselves were not inherently threatening or disruptive objects. Funerary practices involved preparing the corpse at home and viewing the body, and the exemplary display of memento mori or of the criminal dead was common; dead bodies were part of normal experience. From the eighteenth century on, with increasing philosophical scepticism about the soul, medical investigation of the body emphasizing the mechanical function of its dissected parts, and the discipline of the live body as a means of social control (Foucault 1977), the result was an increasingly materialistic approach to death. If people are simply their bodies, without reference to a soul, then the death of the body is the annihilation of the social person. This gives the dead body a new disruptive potential which must be coped with. As Tarlow points out, it is from this moment that we see the rise of specialized ways of distancing ourselves from death in Europe – the beautification of the corpse, the creation of specialist undertakers, a range of new metaphors such as death as a sleep. Combined with declines in mortality in the modern period and with the increasing privatization and individualization of death (Aries 1981), these attitudes have helped to marginalize death from normal experience.

Repersonalizing the Dead

We are circling back to Ötzi via the question of how one may react to dead bodies. The gist of the argument above is that in much of Europe and North America social persons are equated with individual, bounded physical bodies. Deriving from this (and in no way universal), the dead body has a narrative of humanness which has to be resolved, a dynamic tension which, if left unaddressed, threatens all bodies. But what occurs when normal assumptions are violated, when the equation of a physical body and a social person collapses?

Social death which is not accompanied by a physical corpse is problematic: we are unable to resolve the death of the social being without the presence of the individual physical body. Such a situation kicks into action specialized institutions. The first reaction is to find the body. For example, bodies have to come home from wars; thirty-five years after the end of the Vietnam war, US governmental organizations are still identifying and repatriating the remains of servicemen killed there. Mass graves of human rights victims are excavated to provide 'closure' to the families of victims, even when it will not help bring perpetrators to justice.[2] If an airliner crashes, even when the authorities know exactly who was on board and that there are no survivors, great pains are nevertheless taken to identify every human remain as belonging to a particular individual. When such measures fail and no body can be provided to mourn and bury, we erect surrogate foci of memory such as cenotaphs and walls of remembrance. Parallel impulses to memorialization are seen in the construction of commemorative foci for Balkan war crimes victims and for deceased organ donors (cf. Petrović-Šteger, this volume; Sharp 2001).

Conversely, what happens when we have a dead body that is not a social person? This can happen both legitimately – the unclaimed body of a pauper in a morgue, for example, or for that matter most archaeological bodies – and illegitimately, as when an unidentified body turns up. This is a deeply problematic situation; there are two common strategies for coping with it. One is to treat the body as a purely biological or material entity; the other is to recognize its absent sociality and to engage specific technologies to reconstruct it. As Petrović-Šteger (this volume) notes, this does not reflect a duality between a purely scientific, biological approach and a focus on social personhood; both are social strategies. The duality between them reflects the underlying ambiguity with which the body is regarded in modern European culture, as a hybrid of anonymous matter and social personhood.

Depersonalizing the body means forcefully denying its humanity and arguing that it is solely a material object. Medical treatment of cadavers and tissues for training and experimentation is surrounded by a rigid etiquette. In medical contexts, for example, cadavers being dissected often have their identifying information kept confidential, only those parts actually being dissected are exposed to view, and anatomical specimens and medical cases discussed in publications enforce a rigid anonymity. Similarly, transplant recipients may be discouraged from knowing the identity of the dead whose organs they have received (cf. Sharp 1995, 2001); autopsy results are couched in formal scientific language which distances them from the daily experience of bodies. A second context in which it is common to encounter dead bodies which are not associated with social persons is archaeology and biological anthropology (cf. Peers, this volume). Whether or not excavated burials can be associated with specific named individuals (and normally they cannot), the fact that they are excavated and studied archaeologically imposes a regime of depersonalization. Skeletons are given numbers or overtly impersonal designations rather than personal names. They are subjected to medical procedures without the considerations of consent, privacy or invasiveness which contextualize such procedures in the living. They are often exhibited in museums as evidence or curiosities, and they can legitimately be put to use as an educational resource. Harm to them is assessed with reference to their usefulness as scientific or educational resources rather than their integrality as an inherent good. They are stored, moved, catalogued and described as things. The practices of archaeological study prescribe an attitude of vague generalized respect towards skeletons, but also pervasively construct the skeleton as an object rather than a person.

The second strategy is the opposite – to reclothe the body in a social persona, an effort of 'deliberate repersonification' (Lambert and McDonald, this volume). A dead body, if not systematically depersonalized, cannot remain a potential person; it is an anomaly exerting a dynamic tension, it must be someone, an individual, an identified and specific being. Its presence, thus, necessarily calls into operation a range of specialized technologies of individuation, the goal of which is to recreate all the attributes felt to be obligatory for a social person. The standard questions in forensic science include the age and sex of the deceased, and description of biological variations such as 'race', pathologies and the individuating marks which make it different from all other bodies and reveal its unique history. What did the dead person eat? What did he or she look like? What activities did he or she carry out? How did he or she meet their end? Note that these apparently transparent questions encode archaeologists', and forensic scientists', own views of the obligatory categories of personhood. For example, they equate sex and gender transparently, they fit chronological age into a normative biographical narrative within which categories such as 'middle age' have meaning, and they assume that the physical body provides the principal basis for a person's individuation. Note also that these questions call into being an extensive array of methodologies developed for no other purpose. These go far beyond the basic methods for the determination of age and sex from the skeleton. Facial reconstruction, whether necessary or accurate, is often carried out principally for its ability to suggest a living person; a face is much more of a person than a skull (Prag and Neave 1997). Increasingly, DNA is regarded as a necessary part of individuation, not only to assign a specific identity to the dead (as in recent attempts to definitively identify Mozart's skull genetically) but as a way of constructing a sense of relatedness to the dead through common ancestry, a modern analogue for craniometry. As a trend, this use of DNA as an oracle of identity mirrors other popular practices such recreational genealogy, which is predicated upon the assumption that the composite genetic history of an individual body equals its social history and reveals something essential about the person who goes with it. The ultimate goal of the 'reconstruction of life from the skeleton' (Kennedy and

Isçan 1989) in archaeology is to assign a name, a history, a social persona. This is a socially constructive exercise in many contexts, for instance in forensic investigation of human rights abuses (see Petrović-Šteger, this volume). Even more generally, however, as the huge popular interest in forensic studies of dead bodies demonstrates, individuation is for us a compellingly necessary but uncertain process, fraught with revelatory tension.

Within the cultural framework in which archaeologists generally work in Europe and North America, what determines a legitimate disciplinary response to a dead body, which strategy to take – whether to depersonalize or individuate unresolved physical dead bodies? There seem to be three factors: time, bodily preservation and politics.

Beyond a certain limit (a century in the UK, for example) bodies are the business of archaeology rather than the police. Even when remains are clearly archaeological, time is important. The comfort horizon in modern Britain seems to be between one and two centuries before the present; somewhere in the nineteenth century it becomes acceptable to study the dead archaeologically. Importantly, this time horizon, which seems both self-evident and convenient, encodes an entire ontology. As noted above, if we believe that personhood is based upon the individual social relations of the living body, after a certain lapse of time, the dead will no longer be socially relevant, as individuals, to anyone living. They can then be safely depersonalized. If the dead participate in social relations which are collective or enmeshed in a different sense of time, they may remain active social beings for much longer, a fact illustrated in some repatriation cases (see Peers, this volume).

The second factor concerns bodily preservation. Leighton's (n.d.) research, reports from recent burial sites such as Spitalfields (Reeves and Adams 1993), and my own knowledge of archaeological practitioners, all suggest that bodies with soft tissues such as skin and hair are much more likely to cause discomfort. Many, perhaps most, archaeologists and biological anthropologists are comfortable working with dry bones but not with fleshed bodies, which are much more likely to spur feelings of loss, sorrow, sympathy, or revulsion, and to be either subjected to a full forensic identification or reburied without study.

Finally, political relatedness is critical. Relations between living people and groups are constructed through their attitudes towards the dead bodies which mediate between them. When the dead are needed to relate people, they may be constructed as common ancestors and enmeshed in social relations similar to those among the living. In situations of conflict, they may be appropriated, destroyed or redefined. Peers (this volume) describes how the display of 'native' remains was part of constructing colonial difference in the nineteenth century, with acts of collecting, museum processing and scientific analysis serving as rites of transition to redefine the remains as material things rather than as persons. Within modern societies, deprivation of personhood in death has traditionally been seen as a consequence of poverty or marginalization (as in burial in an unmarked pauper's grave, or in a nameless mass grave following war crimes). Marginal, poor, criminal, insane or foreign bodies have been preferentially used in situations when a 'purely biological' body was called for, for instance as anatomical specimens. While participants may have seen this as merely an expedient for locating socially accessible bodies for disvalued purposes, it also underlines and perhaps even helps construct power relations between non-stigmatized and stigmatized groups.

To summarize this argument, in a purely biological sense of lacking metabolic processes, a recent body is as dead as an ancient one, and a body with soft tissues is as dead as a fully skeletonized body, but the condition can force our hand in constructing the chain of events for dealing with it. The latter are likely to be systematically depersonalized as the province of archaeology; the former are likely to be systematically reinvested with those characteristics which are felt to constitute the living individual's social personhood.

Writing Ötzi

The Ice Man, the Academics and the Public

We return now to Ötzi. The basic facts are simple and well known (Barfield 1994; De Marinis and Brillante 1998; Höpfel, Platzer and Spindler 1992; Spindler 1994, 1995). In September 1991, hikers in the high Tyrolese Alps found a dead body protruding from a melting mass of glacial ice. Although the body was unclothed, fragments of fur, skin and woven grass clothing found near it represent its clothing, and varied implements were also found with it – an axe, a bow and quiver full of arrows, a backpack frame, a bark canister for carrying fire, and other objects. The initial series of events show institutions and people responding to a challengingly undefined situation. The body was initially interpreted as an historical mountaineer, perhaps recently deceased. The initial recovery of the body, over five days, was chaotic, complicated by bad weather and jurisdictional doubts. After the realization that the body was genuinely very ancient, based on finds of prehistoric artefacts with it, it was assigned confidently to the Early Bronze Age based on the form of the copper axe found with it, a verdict soon overturned when radiocarbon dates placed it a good millennium earlier. There followed various struggles over control of what was clearly a huge piece of academic and economic capital (Fowler 2000).

When the dust had settled, it was clear that the 'Ice Man' was a genuine prehistoric body, naturally mummified, dating to near the end of the fourth millennium BC. The body is one of the earliest mummies known anywhere. It is accompanied by a unique assemblage, a travelling kit of clothing, tools and personal items. Besides an exhaustive research programme focused upon the body and the artefacts, there have been archaeological excavations at the site, and an international boundary commission (which found that the Ice Man was recovered within Italian territory, as redefined in 1919 by the Treaty of Saint-Germain, by less than 100 metres; Bernhardt 1992). Delicate diplomatic negotiations were needed to allow predominantly Austrian research in the initial phase, followed by permanent curation in the staunchly autonomous German-speaking province of the South Tyrol (Italy), rather than swooping the remains off to a centralized museum in Rome; a new, state-of-the-art archaeology museum was built at Bolzano specifically to house the find. Meanwhile, researchers have built up a detailed picture of the Ice Man. Over the last fifteen years, his body and equipment have been subjected to an exhaustive battery of explorations, spanning everything from craniometry to DNA studies to micro-parasitology and experimental replications of his gear. From this remarkable find we have learned much about the period which would never have been accessible in any other way.

The Ice Man's career in the popular media began immediately with his discovery (Spindler 1994). From the very beginning he was front-page news in Europe and America, and he continues to make headlines with each new analysis. Like any high-profile archaeological discovery, he has attracted the lunatic fringe. The three favourite allegations – all without any real basis – have been that he is actually an Egyptian mummy which was planted as a scientific fake; that finds of semen in his anus reveal him to have been a practising homosexual; and that researchers studying him frequently die due to a curse similar to that supposedly dogging Tutankhamen scholars. Yet we should not be misled by the entertaining fecundity of the popular imagination. In fact, the public Ötzi is remarkably similar to the scholarly one. In discussions of Ötzi on non-professional internet sites, for example, one finds mostly accurate rehearsals of knowledge derived from authorized scientific accounts, supplemented by ongoing press releases from reputable academic researchers. When the fake, curse, or homosexual theories are mentioned, it is almost always to point out the lack of evidence for them. Even Reinhart's colourful theory (which has been published in press releases and popular magazine articles but never in a scholarly venue)[3] that the Ice

Man died as a ritual sacrifice to mountain gods has received remarkably little attention from the general public.

This is thus not a case where popular and academic archaeology occupy parallel but distinct universes. The reason, it is clear, is that academic researchers have been focusing virtually exclusively upon a narrow range of questions which are also the ones the public wants answered. The study of Ötzi has been a predictable exercise in the transformation of a physical body into a social person.

A Research Agenda Launched by the Needs of the Body

The Ice Man's mortal spoils are, in Spindler's phrase, 'a pitiful bundle of humanity' (Spindler 1994: 256). The body is recognizably human, and has evoked feelings of both curiosity and pity (see Figure 4.1). This reaction is documented from the start; those involved in the recovery of the body, an effort lasting five days in all, alternated between viewing the face and showing it to the media and visitors and covering it out of feelings of respect for the dead (Zissernig 1992; Fowler 2000). Similarly, even as he formulated a vast research programme and as the University of Innsbruck engaged a public relations firm to deal with media exposure, Spindler debated whether the body should be on display to the public. In a like vein, a theologian at Innsbruck urged that research should be conducted ethically and with reverence for the tragically unburied dead. 'Precisely the fact that the Ice Man from Hauslabjoch is so well preserved and his human features are so clearly visible does call for a certain piety' (Rotter 1992: 26, quoted in Fowler 2000: 163). An editorial in the *Lancet* recently asked, on similar grounds, whether it is time to halt the relentless study of the Ice Man (Sharp 2002). Public information from the South Tyrol Museum echoes these sentiments:

> We are fascinated, astonished, but also strangely touched to meet a witness of our own past. The fate of an individual human being deprives the 'story' of its anonymity – and it comes alive in our imaginations. (Archaeological Museum of the South Tyrol 2006: 3).
>
> In awareness of the fact that this archaeological find could likely also result in heated ethical discussions, great importance was attached to a very restrained form of presentation. Today, the 'Iceman' section of the museum is characterized by a very sober, scientific atmosphere. Graphic presentations and the architecture do not attempt to compete with the display object … . By means of the partitioning of the exhibition room, the museum visitor can decide for himself if he wants to view the mummy or not. (Archaeological Museum of the South Tyrol 2006: 1).

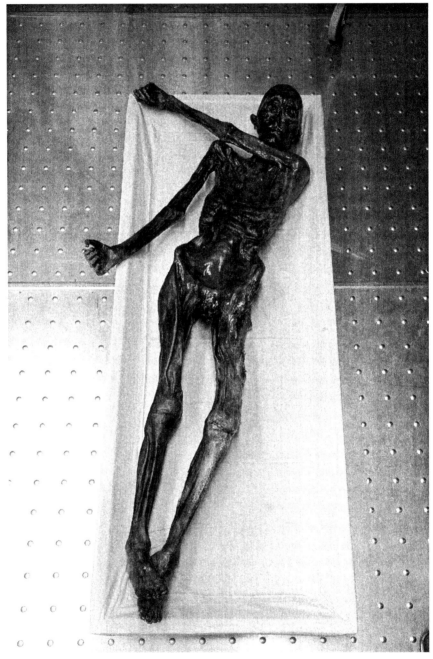

Figure 4.1. The Ice Man's body (photo property of the Photo Archives, Archaeological Museum of the South Tyrol, www.iceman.it, used with permission).

As these reactions suggest, the appearance of the Ice Man's body – recognizably and compellingly human, but divested of the inherent sociality of the body – provoked a tension between alternative reactions. Piety and respect did not suffice to preserve the Ice Man from exhaustive study and public display. However, the exceptional preservation of the body, particularly the presence of soft tissues such as skin, muscle, nails and eyes, was critical in determining how people reacted to the body. Virtually all other human remains contemporary with the Ice Man have been depersonalized archaeologically, handled, studied and published as archaeological burials, and subordinated to standard research agendas focusing on traditional archaeological topics such as funerary ritual, social status or cultural affiliation. Not so the Ice Man. As Fowler perceptively notes, 'From the time of the discovery, the scientists, with very few exceptions, saw not a research project but a *person with a story*' (Fowler 2000: 261, italics in original). The overwhelming goal of research was simply to answer the same questions one would raise in any forensic investigation of an unidentified body, to identify the Ice Man as a unique person and to learn as much about him as an individual as possible. As such, the Ice Man joined a very small, select company of prehistoric media celebrities, some identifiable textually as individuals (e.g. some Egyptian mummies) and others with remarkably preserved soft tissues (e.g. the Lindow Man bog body). Who was Ötzi? Where did he come from? When did he live? How did he live? What did he eat? What did he look like? And, above all, how did he die? Simple as they are, these questions are often asked by the public (Archaeological Museum of the South Tyrol 2006) and have anchored the vast scholarly Ice Man research agenda.

While some are questions important to any archaeological research agenda, others are not, and the focus upon individuation has sometimes distorted or precluded other lines of reasoning. For example, the important examination of the Ice Man's intestinal contents has been marketed as 'The Ice Man's last meal' and interpreted not in relation to a wider Neolithic social context but in relation to the particular sequence of his last few days (Rollo et al. 2002). Similarly, it was quickly noted that the Ice Man's garments were originally sewn by a neat, capable hand, but later repaired by a less expert worker. While this has important implications for issues such as divisions of labour, the observation was simply treated as a clue in a mystery story and shoehorned into particularistic narratives of how the Ice Man died (for instance, in Spindler's account, in which he was cast as a refugee fleeing conflict and forced to mend his own clothes on the road) (Spindler 1994).

Another example arises from Loy's recent argument, not published but widely reported in media and internet sources (and see Archaeological Museum of the South Tyrol 2006), that blood stains on the Ice Man's tunic contain DNA from four or five separate people. Hence, it is proposed, the Ice Man received his fatal wounds in or shortly after a violent melee in which he himself injured or killed others. Suggestive and important data, certainly, and this is indeed a possible interpretation. Yet it is hardly the only one possible. It assumes, for example, that the blood stains must all originate in a single event rather than accumulating over time, that he was not wearing his grass cape over his tunic at the time, that blood is let principally in conflict (rather than therapeutically, ritually or judicially, for example), that the blood came from his adversaries, and so on. The fact that this has been the only interpretation yet put upon this data only shows how tightly Ice Man research has been harnessed to supplying personalizing narratives of his life and death.

Beyond published academic research, Ice Man research has been accompanied by a range of less discursive activities, particularly reconstructions and experimental archaeology. While many archaeologists would probably write these off as merely devices for communicating to the public, dismissing them as ancillary to 'real' research misidentifies their role. For one thing, they often require very extensive research (for instance, the beard and hair in the official reconstruction of Ötzi are based on extensive microscope work), which is undertaken principally to answer the questions founded in the problem of reconstruction. For another, they produce compelling images which are difficult for specialists and public alike to dislodge or unthink and which inform future thinking. Producing such images is in fact a key imaginative step in the process of writing Ötzi.

Thus, the presence of so lifelike a body provokes a research agenda whose main purpose is to repersonalize and resocialize the anonymous and unrelated body. The agenda is deceptively transparent. As with the task of individuation in other contexts such as forensics, there has been very little introspection among Ice Man researchers as to why these are important questions to answer, and to whether this research agenda precludes or distorts other possible ones. Instead, it is simply self-evident that these are the things one needs to know about a body which still contains, in its corporeal integrity, signs of its essential sociality.

Repersonalizing the Ice Man

One of the first facts to be established was the historical age of the body, which was quickly shown to be neither modern nor medieval but prehistoric, pronounced at first Bronze Age and then, when radiocarbon dates became available, Late Neolithic, just at the transition to the Copper Age. This established the jurisdiction over the body (forensic or archaeological) and the possibilities of individuation.

Age and sex followed relatively quickly. The Ice Man was originally posited as a younger adult, but this estimate was eventually revised upward, and, based principally upon his dental wear and bone remodelling, the Ice Man is now thought to have been an older adult, aged probably between forty and fifty, when he died. Interestingly, age can be much more problematic to assess physically than sex, and physical anthropologists always express age estimations as a range of possibility, usually of at least five years, span. In the public information supplied by the South Tyrol Museum press bulletins, however, the Ice Man's age is given as forty-six years, a concreteness which helps visualize him as a specific individual. This age estimate is then framed by placing him within a cultural biography: 'Was Ötzi a grandpa? It's impossible to say if Ötzi was a grandfather. But there is no doubt that he was among the oldest members of his community' (Archaeological Museum of the South Tyrol 2006).

The Ice Man's biological sex, male, was established unequivocally as male genitalia are preserved. While age and sex are important anthropological data, they are also obligatory social categories: one cannot have a Euro-American body which does not have a social age and a gender. At the same time, for both age and sex, there has been a tendency to take the biological at face value. This is most marked when it comes to the question of the Ice Man's gender. The potent combination of a biologically male body and weaponry – his axe, knife and bow – has proved decisive: nobody, as far as I know, has ever seriously queried the assumption that the Ice Man was gendered male. This is defensible archaeologically; in the specific context of later European prehistory in this area, there is strong evidence for dichotomized male and female genders, with weapons forming a diacritic of maleness both in representations and in burials (Whitehouse 1992; Robb 1994; Tafuri 2007). What has not been explored is what maleness meant in the Ice Man's own world. Based on the archaeologists' own cultural assumptions that sex and gender are equivalent and represent simple, lifelong identities, in effect, the designation of maleness has proved a substitute rather than a starting point for discussion of the Ice Man as a gendered body.

Once a body has been aged and gendered, attention turns to biographical details. The Ice Man's life has been investigated through a remarkable ensemble of archaeological, botanical, trace element, isotope, faunal and pollen research (Höpfel, Platzer and Spindler 1992; Spindler 1995; Hoogewerff et al. 2001; Renault-Miskovsky 2001; Holden 2003; Muller et al. 2003; Murphy et al. 2003). The Ice Man probably lived in a small village in the nearby Val Senales, he belonged to a community which herded, farmed and hunted, and he periodically engaged in copper-smithing.

One of the most important thrusts of this research was to understand how the Ice Man could be related to modern populations. How to map the Ice Man onto modern social geography depends on where he was found, where he lived, and biological interpretations of his body. Politically, situating Ötzi involved an intricate and fascinating interplay of interests. The body was found just inside Italian territory, but it was in the South Tyrol, a region which had been Austrian until it was awarded to Italy in the settlement of the First World War. In spite of Mussolini's aggressive Italianization campaign and recent immigration from Italian-speaking Italy, the South Tyrol area remains German-speaking and has close cultural ties to the Austrian Tyrol; it enjoys both a strong identity and considerable self-government as an autonomous region of Italy. Hence there was a strong motive for situating the Ice Man not in terms of nationalities but in terms either of regionalities or of a generic European identity. This has largely been accomplished through the establishment of the special museum at Bolzano (the capital of the South Tyrol), a strongly international research programme, and a rather muted publicity which promotes him, rather like Beethoven's Ode to Joy, mostly as an international emblem of Europeanness. Information distributed by the South Tyrol Museum, thus, bears no mention of Austria or Italy; rather it alternates between announcing that 'Ötzi came from the South Tyrol' so that moving him to the museum brought him 'Home to Bolzano', and presenting him as an ancestor for a vague, international and possibly worldwide 'us' (Archaeological Museum of the South Tyrol 2006). In this context, craniological and DNA studies (Bernhard 1992; Sjøvold 1992; Handt et al. 1994; Rollo et al. 2006) have emphasized principally the affinities between the Ice Man and the modern population both of the region and of Europe. As Sjøvold announces, 'es sehr wahrschienlich ist, dass wir einen richtigen Vorvater vor uns haben' ('it is very probable that we have before us a real forefather'; Sjøvold 1992: 195). While findings from DNA analysis thus far have been completely generic, they form a basis explicitly for forming an emotive bond between Ötzi and modern European people.

Much attention has been paid to the circumstances of his death, often presented as a forensic narrative, a mystery to be solved. He died in late spring or early summer. Spindler's scenario – that the Ice Man was fleeing a disaster such as an attack upon his village – attempted to make sense of his half-finished bow and arrows, his death in an uninhabited zone, and what were then thought to be peri-mortem broken ribs (Spindler 1994; zur Nedden et al. 1994). A violent death became extremely probable with the discovery in 2001 that he had been shot in the back, with an arrowhead still embedded in his body and no sign of healing around the wound (Gostner and Vigl 2002; Murphy et al. 2003). An unhealed cut on his hand suggests fighting recently before death (Nerlich et al. 2003). Most recently, Loy has claimed in the popular media that cuts on the Ice Man's hands and spots of blood from several other individuals on his clothes suggest a death following close combat. Loy's argument has never been published due to his untimely death but has nevertheless gained wide publicity in press releases,[4] and has also gained wide and generally unquestioned currency on the Internet. The question here is not so much whether these scenarios are right or wrong, but rather how they are formulated – as in the canonical genre of crime fiction, it is assumed that the solution will bring disparate evidence into a single story, will show all participants to have acted plausibly according to some standard of rationality, and will represent the culmination and conclusion of the investigation – and whether they are ends in themselves which justify so much intensive academic research.

Then there is personal appearance. This has involved an extensive research programme upon the Ice Man's clothing and physical remains. Skull reconstructions have corrected the postmortem deformation of the skull and microscopic study has distinguished his beard and head hair from animal fibres from his clothing, and revealed how they were cut (Wittig and Wortmann 1992). The basic facts are presented much as one would read them off a driver's licence or police description: ('Blue eyes, dark hair, 1.60 m tall, 50 kg in weight' [Archaeological Museum of the South Tyrol 2006]), together with his shoe size (38) and the comment that, in modern dress, he would look just like us. Reconstruction goes beyond the verbal to the visual in both two-and three-dimensional representations. From Spindler's original book onward, the first serious presentation of the Ice Man, there have been reconstructions of his appearance in virtually every general publication (Barfield 1994; Spindler 1994; De Marinis and Brillante 1998; Fowler 2000). Still more are available upon the Internet and in museums, making him certainly the most reconstructed individual from all of European prehistory. The apogee and standard is the 'official' reconstruction, a three-dimensional, life-size model which, together with the physical body, forms the centrepiece of the Bolzano Museum (see Figure 4.2).

While ostensibly reconstructions are simply illustrative effigies to hang the reconstructed garments upon, they also supply quite detailed and lifelike faces and postures, effectively presenting not the Ice Man's clothing but the Ice Man himself as a person. The reason is straightforward. The Ice Man's body straddles the threshold: it is compellingly human but also so obviously distorted and distanced from a living body as to render it disturbing, repugnant and difficult to relate to positively. In many ways, in fact, it strongly recalls photographs of living and dead inmates of concentration camps, naked, deprived of hair,

Figure 4.2. Reconstruction of the Ice Man, Archaeological Museum of the South Tyrol, Bolzano (photo property of the Photo Archives, Archaeological Museum of the South Tyrol, www.iceman.it, used with permission).

emaciated, distorted in posture, inexpressive. Like them, it is a body stripped of humanity. To be rehumanized and redignified, it must be resupplied with all of these visible components of self. Reconstructions thus take the physical body and reconstitute the social body, in an obligatory act of material completion mirroring the original cultural construction of the body, but inevitably structured around the archaeologists' own logics (see below).

Finally, bodies need names: in modern European and American culture, there is a fundamental distinction between anonymous bodies and named bodies; whether forensically (cf. Petrović-Šteger, this volume), in dealing with museum specimens (cf. Peers, this volume), or archaeologically, bodies without names cannot be social persons. The culmination of forensic individuation, thus, is a positive identification with a specific person. After the hope of identifying the body as a historical hiker proved groundless, a surrogate name had to be created. The body's official scientific designation, according to Spindler (1994: 77), is 'Late Neolithic glacier corpse from the Hauslabjoch, Municipality Schnals (Senales), Province Bolzano/South Tyrol, Italy'. This is clearly a name made up by a committee, and it treads delicately among the complex political sensibilities of the region, but it is so unwieldy that it is virtually never actually used. Probably the most common name in academic work is 'The Ice Man' ('Der Mann in Eis' in German, 'L'Uomo del Ghiaccio' in Italian). While this appears both specific and politically neutral, it is striking that even the scientists studying the body have come to characterize it as an 'Ice Man' rather than a 'glacier mummy': the name not only genders the remains, but also substitutes, in our attention, the embodied person for the mere physical remains.

Competing with 'the Ice Man', and probably far ahead in celebrity name recognition (especially on the Continent), is 'Ötzi'. 'Ötzi' refers to the Ötzthal valley on the Austrian side of the pass where the body was found. As a nickname, 'Ötzi' is more respectful and less politically loaded than alternatives such as the tabloid-sounding 'Frozen Fritz'. Drawing upon a way of coining nicknames common to most European languages, it sounds approachable and even intimate while avoiding jocularity or disrespect. 'Ötzi' was coined within a week of the find by a Viennese journalist, Karl Wendl. Spindler (1994: 77) quotes Wendl: 'This dessicated, horrible corpse must be made more positive, more charming, if it's going to be a good story'. As this shows revealingly, the motive of the name was explicitly to humanize the grotesque relics so that people could relate to them. The tactic was brilliantly successful; the name has become universal in public presentations and is even used by serious academics in professional journals. With both 'The Ice Man' and 'Ötzi' available to use, it becomes a forgotten footnote that the body must have had one or more names and social personae within its own context; our social reconstitution is elided with the unknowable original.

Rationality: The Ice Man's Body as 'Us' Rather Than 'Other'

In making the Ice Man a person, it is difficult to avoid investing his body with familiar rationalities. This occurs in a multitude of small ways where a choice is available between an interpretation which appears reasonable and one which is disquietingly alien. To take a small example, the Ice Man's outer garment was a woven grass cape. From the surviving fragments, this garment could have been worn either hanging from the shoulders, as a cloak or cape, or hanging from the top of his head like a conical tent. Reconstructions, however, invariably follow the former choice, making him appear to wear his garments in a way familiar to Europeans rather than, say, as an indigenous New Guinean. The official, life-size reconstruction which forms the centrepiece of the Bolzano Museum incorporates many such choices (Figure 4.2). An attractive work of art in its own right and generally faithful to the archaeological evidence, this reconstruction nevertheless carefully fashions Ötzi according to Western canons of personhood and portraiture. While the Ice Man's body was found naked, battered and lying down, the reconstructed Ötzi is sculpted standing in an active, alert posture, meeting the visitor's gaze confidently. The body language is familiar; the posture and gestures imply a bodily bearing similar to modern Western Europeans, albeit in different costume. Nor can we overlook the importance of clothing, particularly in a historical tradition in which nakedness and clothedness are loaded with meanings of primitivism or civilization and in which clothing can be powerfully evocative of humanity and personhood. Although he was found nude, he is shown fully clothed with all his garments on, his bow and quiver slung on his back, axe grasped in his right hand ready for use. The effect is to make the body bulkier, stately, imposing. Just as full regalia transforms an ordinary body into a regal body, familiar but powerful, showing Ötzi in this way creates him as a powerful, active figure.

Even subtle details of choreography matter. Body size and form are politicized (Gremillion 2005); obesity may be seen as a class marker in modern Britain, or, to take another example, differences in height make plain the juxtaposition of indigenous peoples and Western ethnographers, soldiers or authorities in colonial-period photographs. Such meanings may be manipulated in statuary. When the message is domination or heroism (as with many Roman leaders, soldiers in modern war monuments, or Classical athletes such as the Riace bronzes), the body is often presented at lifesize or larger, and raised above the viewer. When intimacy or inferiority is conveyed, the body may be made smaller or positioned to receive an encapsulating gaze; reconstructions of pre-human hominids in science museums, for example, often place them on the same footing as the visitor to emphasize their relative smallness, as well as using hair and facial modelling to emphasize their otherness. In contrast to both, modern sculptures which draw attention to the subject as embedded in ordinary life – for example, the sculpted waiting commuters in New York's Port Authority bus terminal – may present exactly life-size sculptures in ordinary poses with no plinth. The presentation of the Ice Man reconstruction represents a specific choice among these possible choreographies. At 155–160 cm, Ötzi was a good 10–20 cm smaller than the average modern European male. In this reconstruction, he has been placed on a low plinth, sculpted to look like a rocky mountain surface, which raises him 15–20 cm off the ground. Critically, this minimizes the difference in height, allowing him to look us directly in the eye. Unnoticed, this is nonetheless significant; it is central to helping him meet modern visitors as an equal, rather than as someone smaller, different, other. It helps place him as one of us, masking rather than emphasizing the bodily difference.

How did the Ice Man's body relate to material culture such as tools and clothing? The modernist taxonomy classifies objects as either functional or decorative, and this point of view has been attributed to the Ice Man as well. Clothing, for example, is renowned in the anthropological literature for both expressing social meanings and transforming bodies (Hansen 2004). The Ice Man affords a unique opportunity to pursue this line of research, as he presents the only actual set of clothing known from his era. In fact, discussions of the Ice Man's clothing have tended to focus upon its functionality, for example in arguing that his grass cloak, skin tunic, fur hat and grass-stuffed shoes may have been designed to cope effectively with high Alpine conditions (Spindler 1994). The Ice Man's bearskin cap is seen as an effective head-warmer; there has been virtually no mention of its possible use as regalia incorporating a direct citation to the largest, most dangerous carnivore in the area. Similarly, whoever made the Ice Man's tunic expended considerable effort in sewing strips of light and dark animal skin together to make a striking striped pattern which moreover resembles that depicted on some representations of human bodies on slightly later stelae (e.g. Lagundo, thought to be of Copper Age date). Yet as far as I know, there has been no discussion of his clothing as visual culture or as self-presentation, aside from superficial and dismissive references to Neolithic fashion. Without minimizing the Ice Man's undoubtedly expert knowledge of his surroundings, the effect has been to familiarize an alien way of dressing by construing it according to modern Western criteria of technological performance; one almost has a picture of Ötzi dressing in prehistoric Gore-Tex much like the mountaineers who found him.

This point can be extended to the Ice Man's technology as well. Technology is embedded in cultural and social logics; yet to modern Europeans and Americans, technology is viewed predominantly as purely functional problem-solving (Pfaffenberger 1992; Dobres 2001). And so the Ice Man: the dominant theme of the analysis of his material culture is its technological appropriateness and efficiency, the accuracy and power of the bow, the careful use of different wood species for different purposes, the survival value of fire and tinder. One rationale for this is obviously to counter the prejudice that prehistoric people were 'primitive' by demonstrating their sophistication and knowledge. However, it is clear that one reason for portraying the Ice Man as Technician is to allow modern visitors to identify with him:

> But according to the museum visitors, it isn't just the chance for a 'face-to-face' meeting with an ancient ancestor from the Chalcolithic Period which stamps itself in their memory. More than anything else, it is the equipment – preserved for the first time – of a Chalcolithic man which they find so enthralling: Frozen together with the man, his clothes, tools, and personal effects have withstood the millennia. Carefully restored and reconstructed by the Roman-Germanic Central Museum in Mainz (Germany), his 'thermal shoes', 'backpack', and the dagger and sheath make it apparent how expediently equipped the Iceman was. It is amazing to note how little difference there is between the Neolithic implements and the standard equipment of a modern mountaineer. Only the materials have undergone a fundamental modernization. (Archaeological Museum of the South Tyrol 2006: 2)

By a similar logic, Ötzi's body itself is invested with a medical rationality. The Ice Man was carrying a piece of dried birch fungus on a cord when he died; he may have been wearing it around his wrist. This kind of fungus is known sometimes to have antibiotic properties (Pöder, Peintner and Pümpel 1992). Predictably, this find has been interpreted almost universally as a kind of first-aid kit, carried in case of illness or injury while travelling, or as a dose against worms (Pöder, Peintner and Pümpel 1992; Capasso 1998; Capasso, La Verghetta and D'Anastasio 1999). It demonstrates his medical knowledge and prudence, his rational care of his body. The medicinal fungus theory has been criticized (Tuñón et al. 1999; Renaut 2004), but remains commonly cited. The point here is not whether or not he may have used the fungus medicinally; it is that the medicinal interpretation has assumed that the Ice Man shares a modern understanding not only of the fungus's properties but also of illness and therapy as bodily phenomena.

Even more strikingly, the Ice Man's remarkable tattoos have transmuted into medical technology. The Ice Man's body is marked with small groups of linear tattoos at the knee, ankle and lower back. There are many reasons why people may be tattooed. Tattoos are generally used to inscribe social meanings permanently upon the body, with references to personal qualities, ritual beliefs, relatedness or status (Gell 1993; Schildkrout 2004). Use of them as medical therapy is known ethnographically in many contexts (Renaut 2004) but is far from the commonest reason to tattoo. However, the Ice Man's tattoos are simple geometrical marks, and they are located in zones generally covered by clothing; this has led researchers to deny them a 'decorative' role and seek a 'functional' role instead. In one version, when X-rayed, the Ice Man happened to show skeletal signs of osteoarthritis in his legs and lower back (zur Nedden and Wicke 1992; Murphy et al. 2003). Osteoarthritis is a degenerative condition found in most prehistoric people dying older than 30–40 years; it is so common at this age as to be normal rather than a pathological variation, and these are the most common skeletal locations to find it. Furthermore, there is no necessary correlation between skeletal manifestations of osteoarthritis, particularly incipient ones, and any experience of pain (Rogers and Waldron 1995); clinicians may find no pain reported by patients with extensive skeletal changes, or vice versa. Nevertheless, it was quickly suggested that the Ice Man's tattoos were intended to address osteoarthritis (Spindler 1994; Sjøvold et al. 1995). A second initial interpretation saw the tattoos as therapeutic cauterizations (Capasso 1993), and later variants related them to acupuncture (Dorfer et al. 1998), saw them as therapy for sciatica (Hegy and Thillaud 2004), or referred to the analgesic value of severing neural fibres while making the tattooing incisions (Archaeological Museum of the South Tyrol 2006). Whatever the exact details, the medical-tattoo interpretation rapidly became the dominant theory, endorsed enthusiastically in Spindler's authoritative book and frequently cited in both academic and popular writings (Renaut 2004). It is striking that all authorities agree that the tattoos must have been therapeutic even when there is no real consensus on how and why. As with the 'antibiotic' fungus, the medicalized therapeutic tattoos give Ötzi's inhabitation of his body a familiar rationality.

Conclusions: The Social Life of Ötzi

The Ice Man has been made into a contemporary social presence, a magnet drawing people to Bolzano to pay homage to the past, an ancestor for Tyroleans, a common forefather of Europeans, an image of how we once were, a reassurance that five thousand years ago, *mutatis mutandis*, we were not so different. He commands considerable resources, he provokes complex and ambiguous emotions, he is widely recognized and invoked, he provides a fixed point in discourses of identity: he exerts agency in fields of social relations around him.

Yet this fate was not the only possible one which could have unfolded when he was found. Against the eroding force of time, which opens a gulf in relatedness much more quickly in European and American cosmology than in many indigenous ones, there was a potent, perhaps irresistible combination of two circumstances. One is the Ice Man's preservation. To anthropologists, archaeologists and the public alike, it is much easier to class wholly skeletal remains as asocial, biological entities which can be emotionally and practically depersonalized and rendered the object of science. This is the normal fate, for example, for almost all people excavated archaeologically, even if they come from contexts far more recent and familiar to us than Ötzi. But non-skeletal corporeality mandates a social presence; soft tissue, particularly faces, commands treatment as an individual person. Other than the Ice Man, European prehistory affords perhaps only two other faces with such potential to be a person. One is the bog body 'Lindow Man' (Brothwell 1986), whose dramatic biographification (Ross and Robins 1989) parallels Ötzi's in many ways.[5] The other is Schliemann's Agamemnon, from the Shaft Graves at Mycenae, hampered by lack of a physical body and, in fact, of any documentable historical veracity, but nonetheless iconic of noble Greek origins in his legendary tradition and resplendent golden mask.

Rather than 'a past with faces' (Tringham 1991), thus, the story here involves the construction of faces with a past. Yet again, human remains can be mobilized as social persons in many ways. The second circumstance shaping the Ice Man's fate is the political context. His case provides a symmetrical counterpoint to that of indigenous remains from colonial situations. In British museums and scholarly institutions, the exhibition and study of Native American, Australian and Maori remains provided the material focal point for the construction of difference, underscoring the otherness of the 'natives' and demonstrating a dominating ability to recategorize them forcefully as material things rather than people (Peers, this volume). The Ice Man's remains have also been made the focal point of a narrative of social relatedness, but in this case it is a narrative not of difference but of similarity, of 'us-ness', of familiarity and identity between him and modern populations.

Giving the Ice Man a name, thus, is anything but coincidental; it is an essential part of repersonalizing this body. The miraculously preserved 'glacier mummy' is a purely material body trace (Figure 4.1). But bodies are social, and when a body is presented, anomalously, outside of social relationships, those finding it have to extemporize, to use the constructional reflexes provided by their society to make good the missing sociality. We can only imagine how someone from another cultural context would have made sense of finding these human remains. Devout Christians may have reburied the body immediately without study, assuming that he must have been Christian and required the proper treatment as such (as in fact has happened occasionally with prehistoric skeletons in Italy: Lo Porto 1972). Medieval people, perhaps, might have associated his miraculous preservation with moral qualities, as with the incorrupt bodies of saints: Ötzi as an unknown holy man. People from cultures with a strong fear of the dead (the Pueblos, for example) may have shunned him, reburied him immediately, or taken steps for spiritual protection from a harmful presence. People with a different view of time and relatedness may have used other grounds than scientific investigation to establish his identity and relation to themselves, understanding him immediately as a close ancestor regardless of the lapse of many millennia (as in some North American repatriation cases such as that of Kennewick Man).

The point of raising these possibilities – all purely hypothetical, yet with well-known parallels – is simply to emphasize that the compelling mandate to identify the body as an individual, repersonalize it, interact with it closely, investigate it scientifically and construct its social persona in ways which make it familiar to us is not a natural or inevitable response to the situation. Rather, it has its origins in the particular understanding of bodies shared by the scientists and their public. Finding Ötzi's body kicked off an intensive cycle of technologies of individuation, originally forensic, then archaeological. The point is not whether this is an accurate creation or not (and in many ways it reflects an immense amount of painstaking and expert work and merits enormous respect), but simply to recognize that it is a creation, our act of supplying what we feel to be the essential dimensions of any body which are missing here. The result, created jointly by archaeologists, heritage administrators, the media and the public, is no longer the anonymous body, the 'glacier mummy' (Figure 4.1), but rather the 'Ice Man' or 'Ötzi' (Figure 4.2), complete with an age, sex, gender, geographical origin, political belonging, appearance, biography and bodily constitution mirroring those familiar to modern Western Europeans.

There remains one important issue to discuss, that of what directions alternative Ice Man research agendas might take. This is beyond the scope of this article, but a few suggestions might be made. If the Ice Man's body is not a universal body but a historically contextualized body, a research agenda which problematizes bodies from other contexts rather than reading them as our own must grapple with prehistoric modes of embodiment. The sociality of the body provides a starting point for this:

the process through which we have imagined Ötzi parallels that through which Neolithic bodies were created by Neolithic people. Bodies are created materially through social relations of procreation, the transfer of bodily substances such as blood, feeding, nurturing, bodily modification, healing, wounding, killing and burying (see Lambert and McDonald, this volume). What does the transformation of the naked, impoverished and isolated body into the richly capable social body tell about the generation of persons? How does his clothing, for example, endow him with an imposing appearance, a range of colours, definition of zones of the body according to a cultural template also visible in statuary, references to animals, skills and social relations? How does it cite colours, qualities, symbols and cosmological qualities? Tools, and the taught and habituated skills they imply, enabled the body to act in ways which were not only technologically but socially necessary; of all the instruments made in the Ice Man's society, why did he carry the selection he did? What capacities were expected of a body of this kind? How does the Ice Man's life and violent death relate to the social meanings of weapons in defining particular bodies which bore them? How can the Ice Man's life and death be related to an interaction of structure and contingency shaped by his material things? The human body is never complete, but always an unfinished project; effectively, the ensemble of things accomplishes a complex and ongoing work of material completion of the social body. This gives us a point of departure for studies of embodied personhood in fourth-millennium Europe.

Acknowledgements

I am grateful to Helen Lambert and Maryon McDonald for careful comments which have improved the argument greatly; to the Leverhulme Trust for funding aspects of this research; and to the Archaeological Museum of the South Tyrol, Bolzano, for permission to reproduce the copyright images used in Figures 4.1 and 4.2.

Notes

1. Interestingly, this opposition underlies theoretical debates about agency; if agency is associated with volitional command of the material, the concept cannot make sense in, and possibly of, any other ontological tradition. Hence the apparent paradox of the agency of dead bodies.
2. In one example which underscores this even more strongly, relatives of some Argentine 'disappeared' victims did not want them identified, as their deaths

would then be considered socially resolved, masking the fact that the criminals responsible for their deaths had not been brought to justice (Z. Crossland, pers. comm.).

3. See for example National Geographic magazine press release dated 15 January 2003 http://news.nationalgeographic.com/news/2002/01/0115_020115iceman.html, accessed 6 June 2008.

4. See for instance 'Iceman Fights Back', Science 22 Aug 2003, vol 301: 1043; the National Geographic press release, based upon emails from Loy to them, dated October 30, 2003 (http://news.nationalgeographic.com/news/2003/10/1030_031030_icemandeath.html, accessed June 6, 2008), and the official Museum information (Archaeological Museum of the South Tyrol 2006).

5. Interestingly, the fairly numerous bog bodies from Denmark, Britain and Ireland, many with very well-preserved faces, have not inspired such biographical effort towards repersonalization as the Ice Man. While this may reflect the particular historical circumstances of the moment of their finding, it may also be due to the very unusual manner of their death and deposition, which is often ascribed to ritual killing, possibly of stigmatized individuals. In reconstruction of both his life and death, the Ice Man, in contrast, has generally been essentialized as a prehistoric Everyman, representing in himself an entire epoch and thus affording much greater ancestor potential.

References

Archaeological Museum of the South Tyrol. 2006. *The 'Tyrolean Iceman' – the Importance of the Find and the Exhibition in the Archaeological Museum (public information sheet)*. Bolzano: Archaeological Museum of the South Tyrol.

——— 2006. *Frequently Asked Questions about 'Ötzi' the Iceman* (public information sheet). Bolzano: Archaeological Museum of the South Tyrol.

Aries, P. 1981. *The Hour of Our Death*. New York: Knopf.

Barfield, L. 1994. 'The Iceman Reviewed', *Antiquity* 68: 10–26.

Bernhard, W. 1992. 'Vergleichende Untersuchungen zur Anthropologie des Mannes von Hauslabjoch', in F. Höpfel, W. Platzer and K. Spindler (eds), *Der Mann im Eis: bericht über das Internationale Symposium 1992 in Innsbruck*, Vol. 1. Innsbruck: Eigenverlag der, universität Innsbruck, pp. 163–87.

Bernhardt, B. 1992. 'Zur frage der Staatsgrenze', in F. Höpfel, W. Platzer, and K. Spindler (eds), *Der Mann im Eis: bericht über das Internationale Symposium 1992 in Innsbruck*, Vol. 1. Innsbruck: Eigenverlag der universität Innsbruck, pp. 66–80.

Brothwell, D. 1986. *The Bog Man and the Archaeology of People*. Cambridge: Harvard University Press.

Capasso, L. 1993. 'A Preliminary Report on the Tattoos of the Val Senales Mummy (Tyrol, Neolithic)', *Journal of Paleopathology* 5: 173–82.

——— 1998. '5300 Years Ago, the Ice Man Used Natural Laxatives and Antibiotics', *Lancet* 352: 1864–64.

Capasso, L., M. La Verghetta and R. D'Anastasio. 1999, 'L'homme du Similaun: une synthèse anthropologique et paléthnologique', *L'Anthropologie* 103: 447–70.

De Marinis, R.C. and G. Brillante. 1998. *La Mummia del Similaun: Ötzi, l'Uomo Venuto del Ghiaccio*. Venice: Marsilio.

Dobres, M.-A. 2001. *Technology and Social Agency: Outlining a Practice Framework for Archaeology*. Oxford: Blackwell.

Dorfer, L., K. Spindler, F. Bahr, E. Egarter-Vigi and G. Dohr. 1998. '5200-Year-Old Acupuncture in Central Europe?', *Science* 282: 242–43.

Foucault, M. 1977. *Discipline and Punish: The Birth of the Prison*. London: Allen Lane.

Fowler, B. 2000. *Iceman: Uncovering the Life and Times of a Prehistoric Man found in an Alpine Glacier*. Chicago: University of Chicago Press.

Gell, A. 1993. *Wrapping in Images: Tattooing in Polynesia*. Oxford: Clarendon.

———— 1998. *Art and Agency: An Anthropological Theory*. Oxford: Clarendon.

Gostner, P. and E.E. Vigl. 2002. 'INSIGHT: Report of Radiological-forensic Findings on the Iceman', *Journal of Archaeological Science* 29: 323–26.

Gremillion, H. 2005. 'The Cultural Politics of Body Size', *Annual Review of Anthropology* 34: 13–32.

Grosz, E. 1994. *Volatile Bodies*. Bloomington: Indiana University Press.

Handt, O., M. Richards, M. Trommsdorff, C. Kilger, J. Simanainen, O. Georgiev, K. Bauer, A. Stone, R. Hedges, W. Schaffner, G. Utermann, B. Sykes and S. Paabo. 1994. 'Molecular-Genetic Analyses of the Tyrolean Ice Man', *Science* 264: 1775–78.

Hansen, K.T. 2004. 'The World in Dress: Anthropological Perspectives on Clothing, Fashion, and Culture', *Annual Review of Anthropology* 33: 369–92.

Hegy, P. and P.L. Thillaud. 2004. 'Ötzi's Tattoos on Reflexion: a New Diagnostic Hypothesis', *L'anthropologie* 108: 107–9.

Holden, C. 2003. 'Forensic Geochemistry – Isotopic Data Pinpoint Iceman's Origins', *Science* 302: 759–61.

Hoogewerff, J., W. Papesch, M. Kralik, M. Berner, P. Vroon, H. Miesbauer, O. Gaber, K.H. Künzel and J. Kleinjans. 2001. 'The Last Domicile of the Iceman from Hauslabjoch: A Geochemical Approach Using Sr, C and O Isotopes and Trace Element Signatures', *Journal of Archaeological Science* 28: 983–89.

Höpfel, F., W. Platzer and K. Spindler. 1992. *Der Mann in Eis: Bericht über das Internationale Symposium 1992 in Innsbruck*, Vol. 1. Innsbruck: Eigenverlag der Universität Innsbruck.

Kaufman, S.R. and L.M. Morgan. 2005. 'The Anthropology of the Beginnings and Ends of Life', *Annual Review of Anthropology* 34: 317–41.

Kennedy, K. and M. Isçan. 1989. *Reconstruction of Life from the Skeleton*. New York: Alan R. Liss.

Leighton, M. n.d. 'Personified Objects or Objectified People: What Are Human Remains for Archaeologists?', Unpublished manuscript.

Lo Porto, F. 1972. 'La tomba neolitica con idola di pietra di Arnesano', *Rivista di Scienze Preistoriche* 27: 357–72.

Muller, W., H. Fricke, A.N. Halliday, M.T. McCulloch and J.A. Wartho. 2003. 'Origin and migration of the Alpine Iceman', *Science* 302: 862–66.

Murphy, W.A., D. zur Nedden, P. Gostner, R. Knapp, W. Recheis and H. Seidler. 2003. 'The Iceman: Discovery and imaging', *Radiology* 226: 614–29.

zur Nedden, D. and K. Wicke. 1992. 'Der Eismann aus der Sicht der radiologischen und computretomographischen Daten', in F. Höpfel, W. Platzer and K. Spindler (eds), *Der Mann im Eis: bericht über das Internationale Symposium 1992 in Innsbruck*, Vol. 1. Innsbruck: Eigenverlag der universität Innsbruck, pp. 131–48.

zur Nedden, D., K. Wicke, R. Knapp, H. Seidler, H. Wilfing, G. Weber, K. Spindler, W.A. Murphy, G. Hauser and W. Platzer. 1994. 'New Findings On The Tyrolean Ice Man – Archaeological And Ct-Body Analysis Suggest Personal Disaster Before Death', *Journal of Archaeological Science* 21: 809–18.

Nerlich, A.G., B. Bachmeier, A. Zink, S. Thalhammer and E. Egarter-Vigi. 2003. 'Ötzi had a wound on his right hand', *Lancet* 362: 334–34.

Pfaffenberger, B. 1992. 'Social Anthropology of Technology', *Annual Review of Anthropology* 21: 491–16.

Pöder, R., U. Peintner and T. Pümpel. 1992. 'Mykologische Untersuchungen an den Pilz-Beifunded der Gletschermumie vom Hauslabjoch', in F. Höpfel, W. Platzer and K. Spindler (eds), *Der Mann im Eis: bericht über das Internationale Symposium 1992 in Innsbruck*, Vol. 1. Innsbruck: Eigenverlag der universität Innsbruck, pp. 313–20.

Prag, J. and R. Neave. 1997. *Making Faces: Using Forensic and Archaeological Evidence*. London: British Museum.

Reeves, J. and M. Adams. 1993. *The Spitalfields Project. Vol. 1, The Archaeology: across the Styx*. York: Council for British Archaeology.

Renault-Miskovsky, J. 2001. 'The Iceman and his Natural Environment. Paleobotanical Results', *L'anthropologie* 105: 639–40.

Renaut, L. 2004. 'Les tatouages d'Ötzi et la petite chirurgie traditionnelle'. *L'anthropologie* 108: 69–105.

Robb, J.E. 1994. 'Gender Contradictions: Moral Coalitions and Inequality in Prehistoric Italy', *Journal of European Archaeology* 2: 20–49.

Rogers, J. and A. Waldron. 1995. *A Field Guide to Joint Disease in Archaeology*. New York: Wiley.

Rollo, F., L. Ermini, S. Luciani, I. Marota, C. Olivieri and D. Luiselli. 2006. 'Fine Characterization of the Iceman's mtDNA Haplogroup', *American Journal Of Physical Anthropology* 130: 557–64.

Rollo, F., M. Ubaldi, L. Ermini and I. Marota. 2002. 'Ötzi's Last Meals: DNA Analysis of the Intestinal Content of the Neolithic Glacier Mummy from the Alps', *Proceedings Of The National Academy Of Sciences Of The United States Of America* 99: 12594–99.

Ross, A. and D. Robins. 1989. *The Life and Death of a Druid Prince: Story of an Archaeological Sensation*. London: Rider.

Rotter, H. 1992. 'Ethische Aspekte zum Thema', in F. Höpfel, W. Platzer and K. Spindler (eds), *Der Mann im Eis: bericht über das Internationale Symposium 1992 in Innsbruck*, Vol. 1. Innsbruck: Eigenverlag der universität Innsbruck, pp. 24–28.

Schildkrout, E. 2004. 'Inscribing the Body', *Annual Review Of Anthropology* 33: 319–44.

Sharp, D. 2002. 'Time to Leave Ötzi Alone?', *Lancet* 360: 1530.

Sharp, L. 1995. 'Organ Transplantation as a Transformative Experience: Anthropological Insights into the Restructuring of the Self', *Medical Anthropology Quarterly* 9(3): 357–89.

——— 2001. 'Commodified Kin: Death, Mourning and Competing Claims on the Bodies of Organ Donors in the United States', *American Anthropologist* 103: 1–21.

Shilling, C. 2003. *The Body and Social Theory*, 2nd edition. London: Sage Publications.

Sjøvold, T. 1992. 'Einige Statistische Fragestellungen bei der Untersuchung des Mannes vom Hauslabjoch', in F. Höpfel, W. Platzer and K. Spindler (eds), *Der Mann im Eis: bericht über das Internationale Symposium 1992 in Innsbruck, Vol. 1.* Innsbruck: Eigenverlag der universität Innsbruck, pp. 188–97.

Sjøvold, T., W. Bernhard, O. Gaber, K.H. Künzel, W. Platzer, and H. Unterdorfer. 1995. 'Verteilung und Größe der Tätowierungen am Eismann von Hauslabjoch', in K. Spindler (ed.), *Der Mann im Eis: Neue Funde un Ergebnisse, Vol. 2.* Vienna: Springer, pp. 279–86.

Spindler, K. 1994. *The Man in the Ice.* London: Weidenfeld and Nicolson.

——— (ed.). 1995. *Der Mann im Eis: Neue Funde un Ergebnisse, Vol. 2.* Vienna: Springer.

Strathern, A. 1996. *Body Thoughts.* Ann Arbor: University of Michigan Press.

Strathern, M. 1988. *The Gender of the Gift: Problems with Women and Problems with Society in Melanesia.* Berkeley: University of California Press.

Tafuri, M. 1007. *La necropoli di Remedello di Sotto: archeologia e antropologia.* Università 'La Sapienza' di Roma.

Tarlow, S. 1999. *Bereavement and Commemoration: An Archaeology of Mortality.* Oxford: Blackwell.

Tringham, R. 1991. 'Households with Faces: the Challenge of Gender in Prehistoric Architectural Remains', in J. Gero and M. Conkey (eds), *Engendering Archaeology.* Oxford: Basil Blackwell, pp. 93–131.

Tunón, H., I. Svanberg, R. Pöder and U. Peintner. 1999. 'Laxatives and the Ice Man', *Lancet* 353: 925–26.

Van Wolputte, S. 2004. 'Hang on to Your Self: Of Bodies, Embodiment, and Selves', *Annual Review of Anthropology* 33: 251–69.

Whitehouse, R. 1992. 'Tools the Manmaker: the Cultural Construction of Gender n Italian Prehistory', *Accordia Research Papers* 3: 41–53.

Williams, H. 2004. 'Death Warmed Up – The Agency of Bodies and Bones in Early Anglo-Saxon Cremation Rites', *Journal of Material Culture* 9: 263–91.

Wittig, M. and G. Wortmann. 1992. 'Untersuchungen an Haaren aus den Begleitfunden des Esimannes vom Hauslabjoch: Vorläufige Ergebnisse', in F. Höpfel, W. Platzer and K. Spindler (eds), *Der Mann im Eis: bericht über das Internationale Symposium 1992 in Innsbruck, Vol. 1.* Innsbruck: Eigenverlag der universität Innsbruck, pp. 273–98.

Zissernig, E. 1992. 'Der Mann vom Hauslabjoch: von der Entdeckung bis zur Bergung', in F. Höpfel, W. Platzer and K. Spindler (eds), *Der Mann im Eis: bericht über das Internationale Symposium 1992 in Innsbruck, Vol. 1.* Innsbruck: Eigenverlag der universität Innsbruck, pp. 234–44.

Chapter 5

BODIES IN PERSPECTIVE: A CRITIQUE OF THE EMBODIMENT PARADIGM FROM THE POINT OF VIEW OF AMAZONIAN ETHNOGRAPHY

Aparecida Vilaça

The primary aim of this chapter is to generate a dialogue between Amazonian or Amerindian ethnography and the literature produced on the theme of the body, especially that centred on embodiment – a notion that forms the basis for most current work on the body. More specifically, I wish to address one of the questions raised by the workshop that preceded this volume: Is 'embodiment' still useful analytically and cross-culturally?[1]

Americanist Ethnography and the Anthropology of the Body

Americanist[2] authors have drawn our attention since the 1970s to the centrality of the body in defining – and differentiating – persons and social groups, as well as the intense use of the body surface – perforated, painted, tattooed and decorated – in the circulation of values. Indeed this theme helped liberate Americanist ethnology from the contemporary exogenous models used in wider anthropology and centred on notions of descent and corporate groups, which had the counter-effect of making the continent's societies appear amorphous, lacking in structuring principles.[3]

Interestingly, the attention Americanists began to show for the body arose independently of the general interest in the theme which started to awaken in anthropology around the same time – related, according to Latour (2004: 227) and the Introduction to this volume, to the emergence of feminism, the growth of science studies, Foucault's work and the expansion of the bioindustry. We could also add to this list: a general rejection of abstract categories and, as some authors put it, 'mentalist patterns of inquiry and explanation' (Strathern 1996: 198; see also Schildkrout 2004: 320), resulting in the favouring of 'a new materialism ... concerned with the domain of lived experience ...'(Lambek and Strathern 1998: 5).

It is likewise interesting to note that a dialogue between Amerindian ethnology and this literature only took off much later – and then in one-way fashion, as some Americanists perceived that the new paradigms emerging from these questions could be applied to their empirical material. But what passed unperceived was precisely the originality of Amerindian conceptions of the body, revealed in all their complexity in more recent work, especially in the studies of Viveiros de Castro and a number of other collaborators developing the theory of perspectivism from the 1990s onwards.[4]

Until this point, the emphasis had been on the holistic aspect of these conceptions: on the intricate relationship between body/soul/mind, as well as on the fabricated nature of bodies, apparently turning upside-down Euro-American genetic conceptions, which had presumed a relatively fixed body determined from the moment of conception. These characteristics had the effect of assimilating Amazonian notions of the body to those being revealed elsewhere and everywhere: not just or even mainly among native peoples, but especially in modern industrial societies, also found to be inhabited by bodies embodied with minds, feelings and affects, completely distinct from the former object-bodies, whose existence was now judged to stem from views distorted by Cartesian presuppositions. Here I refer especially to those theories derived from studies of illness and suffering in medical anthropology that look to deconstruct what are assumed to be exclusively Western dichotomies, such as nature/culture and body/soul, thereby overcoming the great ontological divide between the West and the Rest (see Lock and Scheper-Hughes 1987: 14, 22, 28 and passim; Csordas 1990: 31 and passim; Lock 1993: 136; A. Strathern 1996: 8, 41–62).[5]

As we shall see, the perspectivist reading allows us to discern original properties in this same Amazonian empirical material, enabling the emergence of a body whose central feature is no longer its mindful aspects (although these are not denied) but its capacity to differentiate types of subjects within a universe far transcending the limits of what we conceive as (to be) human. These different bodies do not afford specific views localized within a given (single) universe, but inhabit different and incommensurate universes.

Embodiment

Embodiment, following its use in the last two decades of sociological and anthropological literature on the body, has more than one meaning. In its most commonly accepted sense, the term fixes on its verbal derivation, 'to embody', as shown in a article entitled 'Identity as an Embodied Event' (Budgeon 2003), on the 'embodied experiences of young women' (ibid: 35).[6] This type of argument is based on a specific concept of the person as an individual, constituted by physical attributes which are at the same time psychological and moral.[7] This mindful body – the same one described by Lock and Scheper-Hughes in a well-known article from the 1980s (1987) – is the one providing the basis for intersubjective experiences (Csordas 1999: 181).[8]

The more restricted use of the term 'embodiment' is one that proposes attending to the modes of action and perception of this embodied person as a way of developing a paradigm or methodological tool for analyzing 'culture' (see Jackson 1983; Csordas 1990,). As Csordas (1990: 5; 1999: 187) explains, the idea here is not to propose a new agenda of themes to be studied, but an analysis of the same phenomena from a perspective capable of revealing dimensions obscured by cognitivist and representationalist analyses (see too Lambek and Strathern 1998:13; Reisher and Kathryn 2004: 307).

The foundations of the embodiment paradigm derive from phenomenology, especially the work of Merleau-Ponty. Another important point of reference for this paradigm is Bourdieu's praxeology, which forms a critique of the abstract precepts of structuralism, and was strongly inspired by phenomenological philosophy and by the concept of *habitus* found in Marcel Mauss's 'Body Techniques' (Mauss 1985). Bourdieu's main argument is that structures do not exist in an abstract form, but are embodied, made real – and even transformed – through the action of human subjects (Bourdieu 1972).[9] As his philosophy constitutes the foundations for almost all recent theories concerning the body, including Bourdieu's, I shall focus on Merleau-Ponty, although my reading of his work is limited to specific questions, posed not by the work itself, but by the form in which his ideas have been absorbed and used by proponents of the theory of 'embodiment' he helped initiate.

With Merleau-Ponty's work, the focus of analysis shifts from representational processes, more mental than corporal, to sensible processes of perception on the part of a specific body (Merleau-Ponty 1999: 122). Merleau-Ponty proposes (1980: 247) 'an ontological rehabilitation of the sensible'. In his words: 'The body is no less, but also no more, than the object's condition of possibility' (1980: 254); 'All knowledge, all objective thinking, lives because of the inaugural fact that I feel' (1980: 247–48); 'I am this animal of perceptions and movements that one calls a body' (1980: 248). Following Csordas (1990: 35), this involves 'a return to this level of real, primordial experience', the raw material of perception. Instead of a substrate for cultural symbols, the body is approached as a source of metaphors that constitute culture.[10]

Hence, by paying attention to body attitudes in the possession sessions in a Charismatic Christian church in the USA, for example, Csordas is able to show how the subject's relationship with demons is better expressed by the control-release pair than the interiority/exteriority pair found in analyses focused on the representation of pre-existent cultural objects; in this case, demons.[11] In Csordas's words (1990: 16):

> the language of control/release appears to have as much or greater experiential immediacy. The healer stresses 'release' from the bondage to the evil spirit over expulsion … Bringing to the fore the rather Foucauldian metaphor of bondage points to the concretely embodied preobjective state of the afflicted rather than to the conventionally expressed invasive action of the demonic object.

For this author, the key concept 'allow[ing] us to study the embodied process of perception' is that of 'preobjective' (1990: 9). Returning to Bourdieu's notion of *habitus*, he observes that preobjective is not precultural. In the ethnographic example cited above, we should note that 'control … is a pervasive theme in the North American cultural context …' (ibid. 16).[12] Since culture as a system of representations only exists in embodied form, paying attention to corporal manifestations can help reveal the way in which a determined set of cultural predispositions is actualized in specific contexts – the result of these actualizations being a given objectification/interpretation of the phenomenon. As these authors suggest, this primarily involves a shift in focus: perception is not perceived as the opposite of representation, but its dialogical counterpart (Csordas 1999: 183).[13]

Amazonia

The immediate interest of the notion of embodiment for analysis of Amazonian material lies in the apparent close fit between the phenomenological concept of the body and the empirical data. Here too we find bodies indissociable from minds and affects, and comprising both the seat of perception and the substrate of intersubjective relations. The differences between bodies imply different perceptions, which constitute objects in a particular way, meaning that basing our analysis on the idea of a representation of a given object also leads to a distorted comprehension of phenomena. However, while the phenomenological approach focuses on our concept of the human (albeit not necessarily on the scientific or genetic concept),[14] the Amazonian material obliges us to take as our starting point an extended notion of the human: a notion which comprises a series of beings, including various types of animals, and which is defined above all as a position – an ephemeral vantage point, the temporary outcome of a complex play of perspectives. The phenomenological approach is still based on the Euro-American conception of the person as an individual, which bears no resemblance to the Amazonian conception; the latter, however, does share close similarities with the 'dividual' as described by Marilyn Strathern (and others) in relation to Melanesia.[15]

I turn then to the ethnographic data, concentrating on the Wari', who are Indians living in the Southeast of Amazonia and speakers of a Txapakuran language (see Vilaça 1992, 2006; and Conklin 2001 for comprehensive ethnographic accounts).

For the Wari', the body, *kwere-*, defines a person's way of being. What a person likes and the particular way she acts is determined by the fact 'their body is like that'. The same applies to animals: the peccary wanders in bands because its body is like that. Here we encounter a concept of the body which, though equally designating body substance, includes thinking (in the heart), affects and memory.

Not only the Wari' know themselves to be human: various kinds of animals do so too, including the jaguar, the white-lipped peccary, the tapir, the collared peccary, the capuchin monkey, and so on. This means that these animals perceive their bodies as typically human – although they are not perceived as such by the Wari' and by other kinds of animals – and that they act as humans, possessing what we would call human culture: they live in houses, raise families, hold festivals awash in fermented drinks, embark on hunting trips etc.

The difference between their bodies does not imply, therefore, a culturally differentiated perception of a given universe, as in our relativism, but a particular constitution of this universe – or of this objective world. Thus, what for the Wari' is a cavern, for the jaguar is a house; what for them is blood, for the jaguar is fermented drink; what for them is rotten meat, for vultures is roast meat. Viveiros de Castro (1996, 1998a, 1998b) has dubbed this type of ontology 'perspectivist', or 'multinaturalist', in opposition to our multiculturalism: instead of the same nature and multiple cultures, Amerindians posit the same culture and diverse natures. I shall return to this point later.

Let us turn to the 'dividual' constitution of the person. The Wari' define as human or potentially human all beings possessing *jam,* a term which I have translated elsewhere as soul or spirit. However, unlike other Amazonian peoples, they do not conceive any necessary relation between *jam-* and a vital principle. There are living beings without *jam-*, such as spider monkeys, for example, who lost it after they had stolen some Wari' women (see Vilaça 1992). As a matter of fact, we could say that no living being, when acting in an ordinary manner, has *jam-*. For the Wari', *jam-* implies the capacity to *jamu*, a verb which means to transform, especially in the sense of an extraordinary action. Hence, when people say that a particular animal *jamu*-ed, they mean it acted as a human, shooting and killing a Wari' (an event which appears to Wari' eyes as the victim's sickening and death), or capturing her and turning her into their companion, as in a case I shall discuss later. Similarly, the shaman *jamu*-s when he acts together with his animal partners, perceiving them and being perceived by them as a similar. *Jamu* therefore indicates a capacity to change affects and to adopt other habits, thus enabling the person to be perceived as a similar by other types of beings. This focus on a metamorphic capacity as a central feature of humanity is not exclusive to the Wari' and other Amazonian peoples. Indeed, Ingold makes a similar observation concerning the Ojibwa: 'this capacity of metamorphosis is one of the key aspects of being a person' (2000: 91).[16]

It seems evident that we cannot speak of the body without speaking of the soul, as various Americanists have shown (see Conklin 1996: 375). However, the reason, at least for the Wari' with whom I have been working, seems to be not that the soul gives this body feelings, thoughts and consciousness, but that it gives it *instability*. This is conceived as a capacity typical to humanity which must be controlled since transformation may always be the result of the agency of other subjects rather than ego's (such as the processes of illness conceived by the Wari' as the capture of the soul by animals wishing to make the victim into kin: see Vilaça 2002). This idea of controlling an intrinsic (or even innate) capacity for transformation is present in many Amazonian ethnographies, taking the form of procedures – primarily prophylactic or healing – for impeding the soul's departure or for fixing the soul within the body (Rivière 1974: 431–33).

For the Wari', the soul's relationship to the body is at once symmetric and asymmetric. Considered as a capacity, that is, the potential to adopt an indefinite number of body forms, the soul's relation to the body is equivalent to the relation between the single and the multiple – hence asymmetric. However, since this capacity is always, from an outsider's perspective, actualized as a specific body we can also state that the soul is symmetrical to the body.

Adapting a concept used by Marilyn Strathern (1988) in her analysis of Melanesian ethnography, we could say that while Melanesia reveals dividuals conceived as male and female (Strathern 1988), in Amazonia we are faced with dividuals conceived as human and non-human (or body and soul). But, as in the Melanesian case, when we consider that 'gender difference is not trivial … the crucial difference was that between same-sex and cross-sex relations' (Strathern 1999: 253; 2001), the concept of dividual carries within it a latent asymmetry. Rephrasing my comments[18] in the preceding paragraph on the symmetrical and asymmetrical aspects of the body/soul relation, if the soul is another body, or a body seen from the perspective of the Other, it is also a capacity (or an adjective) in opposition to the body as a realization (or a substantive). Thus we have a pair composed on one hand by a single term (the body) and on the other by an infinite multiplicity of terms (the soul/multiple bodies). We can also perceive this pair in fractal terms: the soul is always decomposable into a body aspect and a soul aspect, as every body – or every realization of the soul – has itself a soul which guarantees its capacity to transform (as same-sex relations can contain cross-sex relations in Melanesia).[17]

With this in mind we can comprehend the prophylactic procedures mentioned above as an eclipsing of the soul, just as in Melanesia it is necessary for one of the genders (or one kind of relation) to be eclipsed for a person to be able to enter into a new relationship (Strathern 1988; see too Gell 1999). In other words, the potential for metamorphosis has to be annulled in order for a specific humanity to be defined. Hence, the Wari' insist that healthy and active people do not have a soul (*jam-*).

This is not an easy point to comprehend. During my first periods of research, I was continually puzzled by the contradiction between the resolutely negative responses to my direct questions on whether people have souls, and the affirmation, in a variety of other contexts, of the existence of this soul, its capture by animals and Wari' sorcerers, causing the person's death and transformation – into an animal of the same species as the aggressor in the first case, and a white-lipped peccary in the second. Eventually I understood that, for the Wari', having a soul meant having an active soul, the person being in the midst of a process of transformation. The prophylactic work to which I refer above involves avoiding this type of process, strengthening the person's ties to their group of human kin. This includes the attention paid to the forms of caring and sharing typical between kin, such as providing food and sleeping together, as well as avoiding dangerous contacts with other subjectivities, whether through the shamanic treatment of game (dehumanization) and adherence to food taboos, or the special care taken in relations with strangers, Wari' from other subgroups and potential sorcerers.

Confronted by this fact, it becomes impossible to comprehend what the literature typically refers to as the fabrication of kinship and bodies[18] if we limit the focus of analysis to processes within the local group. The outside is a constitutive part of kinship relations in Amazonia, for the simple reason that these relations are constructed with alterity as their basic reference point. In other words, the process of differentiating bodies encompasses the process of assimilating: it is by differentiating the child from animals that kin endeavour to make his or her body similar to their own, through food, teaching and direct interventions such as tattooing and piercing.

We can also add that this is a conscious process. The Wari' know that animals see them (humans) as *karawa*, prey, or *wijam*, enemies (conceived as a category of prey) and that they see themselves (animals) as *wari'*, people, human beings. The shaman is the ultimate translator. I once observed a shaman trying to convince jaguars that I was not prey but his kin – and therefore also kin to the jaguars, like himself. He wanted to prevent any act of predation and tried to switch the point of view of the jaguars to his own. Luckily for me, he was successful. The outcome of his act was the transformation of my body in the eyes of the jaguar.

I now wish to turn to the effect of other subjects and non-subjects (objects) on the perception of a particular person – and hence on their bodily constitution.

While the phenomenological approach has the merit of relativizing the effectiveness of analyses focused on representation, forcing us to consider the constitution of objects by perceptive subjects, it constitutes these subjects in advance, making them anterior to the set of relations that produce them.[19] The outcome of this analytical mismatch is far from trivial, given that different relational contexts do not simply produce a subject's distinct objectifications of a determined phenomenon, but different bodily constitutions of this subject. This is particularly relevant in a context where the difference between bodies is so significant: in Amazonia, the result of a relational context may be an Indian, or a jaguar, or a tapir, as we shall see in a short while.

In a recent text, Latour (2004) undertakes an analogous critique of phenomenological approaches, specifically, apropos the anteriority of the perceptive subject. This is particularly relevant in our case since he takes as his starting point not an exotic native society, but the Euro-American universe of the sciences and medicine, focusing his critique on analyses based on the embodiment paradigm. Latour proposes a definition of the body as an interface, as something that 'leaves a dynamic trajectory by which we learn to register and become sensitive to what the world is made of' (2004: 206). The example he provides for us to grasp what 'learning to be affected' means is that of a training course for noses. A kit of odours produced by specialists and an instructor produce a transformation in the body of the person and the universe they inhabit. As Latour writes:

someone able to discriminate more and more subtle differences ... is called a 'nose' as if, through practice, she acquired an organ that defined her ability to detect chemical and other differences. Through the training session, she learned to have a nose which allowed her to inhabit a (richly differentiated odoriferous) world. Thus body parts are progressively acquired at the same time as 'world counter-parts' are being registered in a new way. Acquiring a body is thus a progressive enterprise that produces once a sensory medium and a sensitive world ..." I want to contrast it with another model that may become parasitic on my description ... If we use this model, we will find it very difficult to render the learning body dynamic: the subject is 'in there' as a definite essence, and learning is not essential to its becoming; the world is out there, and affecting others is not essential to its essence. (2004: 207, 208)

For Latour, the body should be conceived as an interface of articulations between various types of subjects and objects, without positing any difference between 'natural' objects and 'fabricated' objects. These objects are not the product of perception, as phenomenology asserts: instead, they play a central role in the constitution of this subjectivity which perceives. The idea 'that culture does not reside only in objects and representations, but also in the bodily processes of perception by which those representations come into being' suggested by Csordas (1999: 183) based on his reading of Merleau-Ponty, or of a 'return to the sensuous quality of lived experience' (Strathern 1996: 198) would make no sense for Latour, who claims that, 'The very idea of a "subjective side" is a myth obtained by discounting all the extrasomatic resources ever invented that allow us to be affected by others in different ways' (2004: 225). This seems to me the central point for the critique of the phenomenological perspective which founds the embodiment paradigm: what we understand as factors of the world, subjects and objects, cannot be disassociated from the perceptive subject. There is no pure perception anterior to the interactions between them.

Let us return to Amazonia. This learning to be affected as an essential part of the constitution of a specific body is exemplified in Wari' funerary cannibalism, where affines and consanguines are differentiated by the inability of the latter to eat one of their dead, whom they do not see as a corpse, associated by the Wari' with animal prey, but as people – as a human being, a kinsperson. The essential function of affines eating the corpse in the ritual is to impose their point of view on the dead person's kin, forcing them to recognize the person's death and thereby differentiate themselves from the deceased via a predator-prey opposition, equivalent for the Wari' to the opposition between the living and the dead. Wari' funerary cannibalism is primarily a question of acquiring or imposing a point of view, a perspective. A dead person's consanguines can only adopt this some time later: this occurs in a ritual marking the end of mourning, when the consanguines mourn over roasted animal prey as if it were their dead relative, and, together with the affines who ate the corpse, eat the prey (which the affines explicitly call 'corpse'). The one who eats constitutes him/herself as human. I should add that had the deceased's kin not been affected by the act of cannibalism realized by affines, they would die of sadness (or would even commit suicide), leaving to join the dead kinsperson in the world of the dead, and acquiring, in the same way as the dead kinsperson and the dead in general, an animal body – that of the white-lipped peccary. Their perspective – now as white-lipped peccaries – would have been radically transformed.[20]

Let us look at another example, which sheds light on the effect of this perspectival dynamic on the resulting type of body. In July 2005, doing fieldwork among the Wari', I heard a report from a woman who was herself captured by a jaguar when a child. This was not the first such report I heard, but the case was particularly interesting since it included the intervention of other people who were listening. The girl was catching fish in the almost dry river when her mother called her to go a bit further on where more fish were to be found. They walked for hours and as night fell, her mother suggested they stop to sleep. In the middle of the night, the girl saw a man arrive and have sex with her mother. Frightened, she asked her mother who the man was. Her mother, still underneath the man who was having sex with her, lulled her to sleep in typical Wari' fashion, smacking her bottom lightly. The following day, they started walking again until suddenly the girl started hearing voices calling her and warning her to move away from the jaguar. Her mother then told her she was going off a moment to defecate and failed to come back. The girl's family found her and took her back home. The mother was actually a jaguar.

During her account, a Wari' woman next to me asked the narrator, But didn't you see any spots on her body or a small tail? No, answered the woman. It was my mother, my mother completely. The question has an empirical grounding: the inquiring woman had heard a similar report from her own mother, who had also been kidnapped, but by a jaguar she had seen as her maternal uncle. However, at the moment when she heard the calls of her family who were looking for her, she suddenly noticed strange signs on the body of the uncle who was leading her, precisely spots on his hands and a discreet tail. As he had carried her on his back during the journey, her body was also covered in jaguar hair.

This report is particularly interesting since it involves a set of perspectives at work that result in specific corporeal forms. Applying the embodiment paradigm to this case is useful insofar as it frees us from explanations based on an appearance/essence opposition, since the representation of the jaguar as a mother was not a trick or illusion, but the result of the girl/subject's perception of a specific kinship relation: the kind one has with mothers and not jaguars. At the same time, though, by limiting the analysis to the girl/subject's perspective, it ignores an important aspect of the phenomenon: its outcome. The perspectives of the Wari' – the girl's kin – and the jaguar must also be taken into account. The jaguar wished to take the girl to live with its own folk, since it wanted a relative, a companion. For this reason it acted as a close relative in order to capture her. The Wari' 'recognized' the jaguar's perspective; that is, they knew it was acting in human form (they would say the jaguar *jamii-ed*), although they could not see it in this way. It should be noted that in the woman's account, they did not depart in search of her armed to kill an animal, but ready to talk to a person. They spoke to the jaguar in Wari' language, asking it to leave the girl alone, arguing that they were her true family and that she should stay with them. The jaguar listened and abandoned the girl. This abandonment, added to the proximity of her kin, affected the girl and altered her point of view. The result was the constitution of her body as a human rather than as the body of a jaguar. Likewise, the jaguar's body was (re)constituted as an animal body, no longer human – an outcome which becomes clear in the tale of capture by the maternal uncle, in which the jaguar's animal body is gradually revealed to the onlooking girl in the form of dark spots on the skin and a small tail.

The difference between jaguars and tapirs is interesting. Tapirs are rarely sensitive to (or affected by) the appeals made by the relatives of those they capture. As a result, these people become tapirs: in other words, they start to accompany the tapirs and were they to be recognized by their kin in the future, they would be perceived as tapirs. The rare case of a boy who was freed illustrates the effect of relations on bodies. Even though at home in his village, he began to escape every night to wander with the tapirs. The man who told me this case added that he had gone to visit the boy in his village soon after he returned and saw that not only was his body covered in enormous tapir lice, his hands were a strange shape, more reminiscent of tapir hooves.

These different perspectives, adopted by a particular person (and here I include the jaguar) according to different situations, are the result not only of different sensibilities enacted at a given moment by a body holistically endowed with mind or spirit, like the body posed by phenomenology, but also of a person's intrinsic capacity to transform, arising from the person's double nature, composed of body and spirit, as distinct and dissociable entities. The dual nature of the Wari' girl, eclipsed in daily life by the care taken by her family to avoid undesirable and dangerous contacts with other humanities, was actualized via the relationship with the jaguar.

This brief analysis of the case of capture by the jaguar brings us to the idea of equivocation as a mode of knowledge in perspectival ontologies, an idea recently developed by Viveiros de Castro (2004). He writes:

> The problem for indigenous perspectivism is not therefore one of discovering the common referent (say, the planet Venus) to two different representations (say, Morning Star and Evening Star). On the contrary, it is one of making explicit the equivocation implied in imagining that when the jaguar says 'manioc beer' he is referring to the same thing as us (i.e., a tasty, nutritious and heady brew) ... Therefore, the aim of perspectivist translation ... is not that of finding a 'synonym' (a co-referential representation) in our human conceptual language for the representations that other species of subject use to speak about the same thing. Rather, the aim is to avoid losing sight of the difference concealed within equivocal 'homonyms' ... since we and they are never talking about the same things. (2004: 6–7)[21]

Recognizing distinct perspectives is no trivial matter, since, as we saw, it directly affects the subject's constitution and consequently the world inhabited by her.

The Worlds of Bodies

It remains for us to think of a way of describing the worlds inhabited by these unstable bodies, in a constant process of being affected by other subjects and objects. Here it is worth observing that the multiplicity of perspectives not only affects what people conceive as persons, but also 'non-persons like rocks, water, air or smoke, which appear to possess an existence of their own, a nature irreducible and indifferent to relations' (Lima 2002: 13). As Viveiros de Castro (1998b: 51; see also 2002: 382–87) remarks, 'what seems to be happening in Amerindian perspectivism is that substances named by substantives like "fish", "snake", "hammock" or "canoe" are somehow used as if they were relational pointers ...'. They can be compared to kinship terms: 'You are a father only because there is another person to whom you are a father: fatherhood is a relation ... something would be "fish" only by virtue of someone else to whom the thing in question is a fish' (Viveiros de Castro 1998b: 51). Hence propositions such as 'people are monkeys to jaguars', examples of which abound in Amazonian ethnographies, are 'of the same nature as a proposition such as: "my uncle is grandfather to my son"' (Lima 2002: 15).[22]

This allows us to highlight another fatal inadequacy of the embodiment paradigm, one that relates to the concept of an objective reality intrinsic to the phenomenological approach. Although the latter states that the object does not exist a priori, but is constructed by the embodied perspective, it also anticipates the transcendental sum of an indefinite number of perspectives through which the object can be (presumed to be) reconstructed in its extrinsic reality. In the words of Merleau-Ponty (1964: 15), the object of perception 'is given as the infinite sum of an indefinite series of perspectival views in each of which the object is given, but in none of which it is given exhaustively'. Csordas (1990: 38) underlines this point in stating that 'This perspective does not deny that objects are given; as I have emphasized throughout this essay, the body is in the world from the start'. This echoes a similar observation by Haraway who proposes swapping the relativist view for the recognition of *location* and hence 'accepting the interpretative consequences of being grounded in a particular standpoint – the consequences of relatedness, partial grasp of any situation, and imperfect communication' (1991: 197–98, also cited in Csordas 1999: 180). The same idea of location is expressed by Merleau-Ponty (1980: 247): 'it is really necessary for my body to be enmeshed with the visible world: it owes its power precisely to the fact it possesses a place from which it sees'.

Even Latour's interesting concept of the multiverse, based on the idea that differently constituted bodies (like those with a nose developed to perceive subtle differences between smells) inhabit – and constitute – different universes, (pre)supposes the possibility of unification. In Latour's words: 'the *multiverse* designates the *universe freed from its premature unification* ... This does not mean ... that we abandon unity, since we do not go from one universe to multiple worlds – we still talk about *the* multiverse – but that we don't want a unification which would have been on the cheap and without due process' (2004: 213).

By contrast, in Amazonian perspectivism the different perspectives do not add up; they are neither complementary nor equivalent. Each perspective constitutes a universe apart, a world parallel to other worlds, and these can only be related through the recognition and valorization of equivocation, in the sense given to the term by Viveiros de Castro (2004) in the above citation. In the absence of the idea of a predetermined objective world capable of being totalized at any moment, a complementarity of perspectives becomes impossible. Another essential aspect of the notion of perspective is its constant transformation: the perspectives of different beings are not defined a priori as the character of the species or group. A perspective is above all a position within a specific relational context. This explains why perspectives cannot be said to be equivalent: the Wari' and others do not posit, for example, that the positions of predator and prey are equivalent. In the case of the jaguar cited earlier, the perspective of the Wari' is imposed, through the action, on the perspective of the jaguar, which in this specific context turns into an animal. As Lima observes in relation to Juruna cosmology, 'alien perspectives are not, in theory, less true than the human perspective … however, this does not mean they are equivalent or symmetrical, as they appear to us when we reduce this cosmology to a known world. Given that this is above all a lived world, and given that human existence here appears primarily as a human struggle, the relation between two or more perspectives is necessarily asymmetric' (Lima 2002: 19).

Another example may provide some insight into this relation between bodies and perspectives. In his model of bodies as interfaces, constituted in relation to the world, Latour (2004: 211–12) suggests that the more embodied (interactive, affected) the body, the more ample or more total its perspective. This idea is suggestive, since in Amazonia – or among the Wari' at least – the precise opposite occurs: the *absence* of the body is the only way of obtaining a total perspective. I am not referring to the dead, who acquire the bodies of peccaries or other animals, but to the Christian God, whom the Wari' have learnt about from fundamentalist Protestant missionaries (see Vilaça 1996a, 1996b, 1997). Talking about this God, in their services or to myself, the anthropologist, the Wari' insist on the fact he has no body. It is precisely this characteristic – so alien to the Wari' universe, where everyone and everything has a body – which enables the equally alien possibility of a total perspective. God can see everything and everyone (see Vilaça 2003).

In her contribution to this volume, Strathern reiterates the singularity of this notion of perspectivism by opposing it to the notions of context or perspective (related to 'perspectivalism') used by scientists, for example, in solving bioethical problems surrounding medical research, or issues raised by the transit of body fragments, such as the polemic over the repatriation of the mortal remains of ancestors, analysed in the chapter by Peers. For the scientists involved, the solution can seem to involve placing contexts of knowledge in relationship and making explicit, to the natives, the different perspectives on a particular phenomenon, enabling them to comprehend their own perspective as just one among various. Underlying this practice is the scientists' belief in a shared reality perceived from different perspectives and, consequently, in the possibility of reaching a consensus through negotiation, ideally based on the combination of as many different perspectives as possible. As Strathern remarks, this idea is completely distinct from the notion of perspective found in Amazonian perspectivism, which is much closer to the native models cited in her text (African, Australian and Melanesian). This is no longer an epistemological problem, as the scientists might assume, but an ontological question. Distinct notions of persons and relations are involved – persons who inhabit different worlds, which can never form a whole of any kind. In contrast to Amazonian Indians, as described by Viveiros de Castro (2004), who take equivocation as a positive effect, turning it into a tool of knowledge and action (they know the jaguar has a distinct perspective, in the sense of inhabiting a particular world, understood as the realization of a specific relational context, and act with this in mind the whole time), scientists take the equivocation as a mistake, a failure in comprehension, and look to fix this problem by widening the perspective of the natives – but only the natives, since the global view is presumed to be intrinsic to science. Like the Christian God of the Wari', scientists possess the global point of view, but they lack a body in Latour's sense (2004), which would allow them to understand what distinct perspectives really mean.

Notes

1. This chapter is a modified version of the article 'Chronically Unstable Bodies: Reflexions on Amazonian Corporalities', published in *The Journal of the Royal Anthropological Institute* (2005) 11(3): 445–64. I thank the editors of this volume and Eduardo Viveiros de Castro for their reading and comments on this version.
2. Here I mean specialists in lowland South American indigenous populations.
3. Overing (1977) provides one of the first consistent critiques of applying exogenous models to Amazonia, formulated as a discussion proposal at a session of the Americanists Congress. In Seeger, Da Matta and Viveiros de Castro (1979) these critiques, combined with a reading of Lévi-Strauss's *Mythologiques*, enabled the authors to suggest the centrality of native conceptions of corporeality in comprehending the sociocosmologies of the South American lowlands; see too Viveiros de Castro (2002: 16).
4. See Viveiros de Castro (1996, 1998a, 1998b, 2002); Lima (1996, 2002); Vilaça 1996a, 2002, 2005, 2006).
5. See Pollock (1996: 320; also n. 2) for a critique of the embodiment paradigm as an attempt to overcome mind/body dualism.
6. The author attests that if 'cosmetic surgery allowed the young woman in question to live her embodied self differently then it would be an acceptable choice for her to make' (Budgeon 2003: 46).
7. For A. Strathern (1996: 197), 'What is embodied is always some set of meanings, values, tendencies, orientations, that derive from the sociocultural realm'.
8. See Viveiros de Castro (1999: 1–2) for a critique of the embodiment perspective.
9. Lock (1993: 137) comments on Bourdieu: 'Drawing on a reformulation of Mauss' concept of habitus, Boudieu's theory was explicitly grounded in the repletion of unconscious mundane body practices. Formulated in opposition to Lévi-Strauss, it was designed to overcome a rigid dualism between mental structures and the world of material objects'.

10. Csordas (1994: 17) states that the body should be conceived as 'a generative source of culture rather than as a tabula rasa upon which cultural meaning is inscribed'.

11. See Van Wolputte (2004: 258).

12. For Csordas (1990: 11), 'Bourdieu goes beyond this conception of habitus as a collection of practices, defining it as a system of perduring dispositions which is the unconscious, collectively inculcated principle for the generation and structuring of practices and representations ... the habitus does not generate practices unsystematically or at random because "there is a principle generating and unifying all practices"' (Bourdieu 1977: 124).

13. In the words of Lambek and Strathern (1998: 15), 'The pre-objective is the realm of experience before it becomes fully "cultured" or "enculturated" ... It has a use in analysis because it provides for the *genesis* of phenomena ... But the pre-objective is always regarded also as incomplete. It looks to objectification to its completion ... objectification implies the transmutation of the embodied gesture, the expressed impulse, into a form that can stand as a symbol of values. The virtue of the term pre-objective is to suggest not an evolutionary argument, but an ontological one'.

14. See A. Strathern (1996: 197) for the same critique: 'the concentration on the human body as such may cause us to miss the ethnographic point in cultures in which the body is seen as a part of a wider cosmos'. On the other hand, in a section entitled 'Embodiment' in the book edited by himself and M. Lambeck, *Bodies and Persons* (Lambek and Strathern 1998), the utility of this paradigm is reaffirmed: 'In our view, there is no doubt that the concept of embodiment does indeed provide some strategic advantages over a generalized approach to personhood' (1998: 13).

15. A. Strathern (1996: 189) comments on Csordas: 'It is evident, also, that the level of action that is here being theorized is primarily that of the individual, albeit the culturally situated individual'. See also Battaglia quoted in Wolputte (2004: 261).

16. The Wari' soul, like the Melanesian *mana* or *imunu*, is 'a quality or a set of qualities, rather than a thing' (Williams 1923: 362–63, cited in Lévy-Bruhl [1927] 1996: 5).

17. To help explain this point, we can note that the pure, soul-less body is the corpse, which in fact is generally not called 'body' by Amazonian peoples. Corpses do not transform. See Viveiros de Castro (2000, 2001, 2002: 444) on the Amazonian dividual. On the specific conception of fractals, see Viveiros de Castro (2002: 439–40). See too Kelly Luciani (2001) for the same type of comparisons and for an exploration of the notion of fractality in Amazonia.

18. Here I refer to those works which take kinship to be a product of effective social relations, which focus on the production of the similar bodies that characterize kin (see McCallum 1996 and Carsten 1995), in opposition to the dominant Euro-American view of kinship as a given fact prior to relations.

19. Although the body emerges in the embodiment paradigm as the grounds for everything, it should be noted that at some points Merleau-Ponty appears to emphasize the mutual constitution of bodies, as in this specific passage: 'The

constitution of another does not come after that of the body: both are born together from the original ecstasy. Corporeality to which the primordial thing pertains is above all a corporeality in general' (1980: 254–55). I have yet to find this mentioned in later formulations of embodiment theory. On this point, Latour's perspective, mentioned below, is perhaps the closest to Merleau-Ponty. An important detail needs to be added to this idea of a mutual constitution of bodies: namely, these relations which produce bodies occur between different types of beings and that specific humanities are defined on this basis.

20. On Wari' funerary cannibalism, see Vilaça (1992, 2000).

21. He continues: 'Amerindian ontologies are inherently comparative: they presuppose a comparison between the ways different kinds of bodies "naturally" experience the world as an affectual multiplicity. They are, thus, a kind of inverted anthropology …' (Viveiros de Castro 2004: 7).

22. Strathern (2002) stresses the point made by Viveiros de Castro in claiming that Amerindian perspectivism is 'a matter of ontology not epistemology' (2002: 6; see also this volume). When a son 'inhabits' the world as a sister's son, 'The world is changed not simply because the son comes to see his father through different eyes (his MB's), but because the mother's brother has made a different subject out of him … . It is an ontological switch' (ibid. 26).

References

Bourdieu, P. 1972. *Esquisse d'une théorie de la pratique*. Genève: Librairie Droz.

Budgeon, S. 2003. 'Identity as an Embodied Event', *Body and Society* 9(1): 35–55.

Carsten, J. 1995. 'The Substance of Kinship and the Heat of the Hearth: Feeding, Personhood, and Relatedness among the Malays in Pulau Langkawi', *American Ethnologist* 22(2): 223–41.

Conklin, B.A. 1996. 'Reflections on Amazonian Anthropologies of the Body', *Medical Anthropology Quarterly* 10(3): 373–75.

——— 2001. *Consuming Grief: Compassionate Cannibalism in an Amazonian Society*. Austin: University of Texas Press.

Csordas, T. 1990. 'Embodiment as a Paradigm for Anthropology', *Ethos* 18: 5–47.

——— (ed.). 1994. *Embodiment and Experience: The Existential Ground of Culture and Self*. Cambridge: Cambridge University Press.

——— 1999. 'The Body's Career in Anthropology', in H. Moore (ed.), *Anthropological Theory Today*. Cambridge: Polity Press, pp. 172–205.

Gell, A. 1999. *The Art of Anthropology: Essays and Diagrams*, edited by Eric Hirsch. London: Athlone Press.

Haraway, D. 1991. *Simians, Cyborgs and Women: The Reinvention of Nature*. New York: Routledge.

Ingold, T. 2000. *The Perception of the Environment: Essays on Livelihood, Dwelling and Skill*. London: Routledge.

Jackson, M. 1983. 'Knowledge of the Body', *Man* 18: 327–45.

Kelly Luciani, J. 2001. 'Fractalidade e troca de perspectivas', *Mana. Estudos de Antropologia Social* 7(2): 95–132.

Lambek, M. and A. Strathern (eds). 1998. *Bodies and Persons: Comparative Perspectives from Africa and Melanesia*. Cambridge: Cambridge University Press.

Latour, B. 2004. 'How to Talk about the Body? The Normative Dimension of Social Science Studies', *Body & Society* 10(2–3): 205–29.

Lèvy-Bruhl, L. [1927] 1996. *L'âme primitive*. Paris: Quadrige/PUF.

Lima, T. 1996. 'O dois e o seu múltiplo: reflexões sobre o perspectivismo em uma cosmologia tupi', *Mana. Estudos de Antropologia Social* 2(2): 21–47.

——— 2002. 'O que é um corpo?', *Religião e Sociedade* 22(1): 9–19.

Lock, M. 1993. 'Cultivating the Body: Anthropology and Epistemologies of Bodily Practice and Knowledge', *Annual Review of Anthropology* 22: 133–55.

Lock, M. and N. Scheper-Hughes. 1987. 'The Mindful Body', *Medical Anthropology Quarterly* 1(1): 6–41.

Mauss M. 1985. 'Les techniques du corps', in *Sociologie et Anthropologie*. Paris: Quadrige/PUF, pp. 365–86.

McCallum, C. 1996. 'The Body That Knows: from Cashinahua Epistemology to a Medical Anthropology of Lowland South America', *Medical Anthropology Quarterly* 10(3): 347–72.

Merleau-Ponty, M. 1964. *The Primacy of Perception*, translated by James Edie. Evanston, IL: Northwestern University Press.

——— 1980. 'O filósofo e sua sombra', in *Os Pensadores: Merleau-Ponty*. São Paulo: Abril Cultural.

——— 1999. *Fenomenologia da percepção*, translated by Carlos Alberto Ribeiro de Moura. São Paulo: Martins Fontes.

Overing, Joanna. 1977. 'Orientation for Paper Topics' and 'Comments'. Symposium 'Social Time and Social Space in Lowland South American Societies', *Actes du XLII Congrès International des Américanistes* 2(9–10): 387–94.

Pollock, D. 1996. 'Personhood and Illness among the Kulina', *Medical Anthropology Quarterly* 10(3): 319–41.

Reisher, E. and K. Kathryn. 2004. 'The Body Beautiful: Symbolism and Agency in the Social World', *Annual Review of Anthropology* 33: 297–317.

Rivière, P. 1974. 'The couvade: a problem reborn', *Man* (ns)9 (3): 423–35.

Schildkrout, E. 2004. 'Inscribing the Body', *Annual Review of Anthropology* 33: 319–44.

Seeger, A., Viveiros de Castro, E. and Da Matta, R. 1979. 'A Construção da Pessoa nas Sociedades Indígenas Brasileiras', *Boletim do Museu Nacional* 32: 2–19.

Strathern, A. 1996. *Body Thoughts*. Ann Arbor: University of Michigan Press.

Strathern, M. 1988. *The Gender of the Gift: Problems with Women and Problems with Society in Melanesia*. Berkeley: University of California Press.

——— 1999. *Property, Substance and Effect: Anthropological Essays on Persons and Things*. London and New Brunswick, NJ: The Athlone Press.

——— 2001. 'Same-sex and cross-sex relations: some internal comparisons', in T. Gregor and D.Tuzin (eds), *Gender in Amazonia and Melanesia: An Exploration of the Comparative Method*. Berkeley: University of California Press, pp.221–44.

———— 2002. 'Divided Origins and the Arithmetic of Ownership', University of California Irvine, Critical Theory Institute, Futures of Property and Personhood. Draft copy.

Van Wolputte, S. 2004. 'Hang on to Your Self: Of Bodies, Embodiment, and Selves', *Annual Review of Anthropology* 33: 251–69.

Vilaça, A. 1992. *Comendo como gente. Formas do canibalismo wari'(Pakaa Nova)*. Rio de Janeiro: ANPOCS/Editora da UFRJ.

———— 1996a. 'Quem somos nós. Questões da alteridade no encontro dos Wari'com os Brancos'. Ph.D. Dissertation. PPGAS/Museu Nacional/UFRJ.

———— 1996b. 'Cristãos sem fé: alguns aspectos da conversão dos Wari' (Pakaa Nova)', *Mana: estudos de Antropologia Social* 2(1): 109–37. Republished in Robin Wright (ed.), *Transformando os Deuses. Os múltiplos sentidos da conversão entre os povos indígenas do Brasil*. São Paulo: Editora da Unicamp, 1999, 131–54.

———— 1997. 'Christians without Faith: Some Aspects of the Conversion of the Wari' (Pakaa Nova)', *Ethnos* 62(1–2): 91–115.

———— 2000. 'Relations between Funerary Cannibalism and Warfare Cannibalism: the Question of Predation', *Ethnos* 65(1): 84–106.

———— 2002. 'Making Kin out of Others in Amazonia', *Journal of the Royal Anthropological Institute* (n.s.) 8(2): 347–65.

———— 2003. 'Big Brother Wari': The Effects of the Idea of God in a Perpectivist Cosmology'. *51st International Congress of Americanists 14–18 July 2003*, Santiago (Chile).

———— 2005. 'Chronically Unstable Bodies. Reflection on Amazonian Corporalities', *Journal of the Royal Anthropological Institute* (n.s.) 11(3): 445–64.

———— 2006. *Quem somos nós. Os Wari' encontram os Brancos*. Rio de Janeiro: Editora UFRJ.

Viveiros de Castro, E. 1996. 'Os pronomes cosmológicos e o perspectivismo ameríndio', *Mana. Estudos de Antropologia Social* 2(2): 115–43.

———— 1998a. 'Cosmological Deixis and Amerindian Perspectivism', *Journal of the Royal Anthropological Institute* 4(3): 469–88.

———— 1998b. *Simon Bolívar Lectures*. Department of Social Anthropology, University of Cambridge (UK). MS.

———— 1999. *Le diable dans le corps: à propos de Jacques Galinier, La moitié du monde: corps et cosmos dans le rituel otomi* (P.U.F., 1990). Unpublished manuscript.

———— 2000. 'Atualização e contra-efetuação do virtual na socialidade amazônica: o processo de parentesco', *Ilha* 2(1): 5–46.

———— 2001. 'GUT Feelings about Amazonia: Potential Affinity and the Construction of Sociality', in L. Rival and N. Whitehead (eds), *Beyond the Visible and the Material: The Amerindianization of Society in the Work of Peter Rivière*. Oxford: Oxford University Press, pp.19–43.

———— 2002. *A inconstância da alma selvagem*. São Paulo: Cosac & Naify.

———— 2004. 'Perspectival Anthropology and the Method of Controlled Equivocation', *Tipiti* 2(1): 3–22.

Chapter 6

USING BODIES TO COMMUNICATE

Marilyn Strathern

It is possible to argue endlessly about what one might wish particular concepts to convey; however, when they are already overdetermined by diverse usage, strategies other than persuasion by argument might help. This is true of 'the body' and 'bodies'. Anthropologists could well turn to the circumstances under which these concepts emerge as entitities in people's dealings with one another. Or they could have recourse to what anthropologists have always done and, so to speak, creep up on them unawares. This way one could also, and so to speak again, let them emerge after analysis rather than have them prepared beforehand.

At least, it was thus that I found myself needing to talk about an entity I called body in a situation that on the face of it did not seem about bodies at all. This was at a conference held under the local auspices of the Kenya Medical Research Institute (KEMRI) on the ethnography of medical research in Africa. The situation concerned questions about the transmission of information where current ethical protocols turn on informed consent (e.g. Molyneux, Peshu and Marsh 2004: 2547), and where (in the conference convenor's words) 'the assumption of most medical field researchers is that a well-informed study population will consent to a reasonable study regimen' (Geissler 2004). The notions of context and perspective on which such essentially Euro-American ideas about information rest led me to think of accounts drawn from parts of Africa where these epistemological assumptions may not have the same place. Indeed it is very interesting for my own purposes that Aparecida Vilaça (this volume) should already have raised the question of 'perspective'. For the kind of multiplying perspectives with which she takes issue prompts an analogy with the concept of 'context', an infinitely pluralizing artefact central to much Euro-American epistemology, and why that should also emerge as interesting will I hope be clear by the end.

My position is very close to Vilaça's, although I come to it through the work of other theorists altogether, namely Mol (2002) and Law (2004).

But why dwell on how one arrives at a topic, and why allow it to emerge rather than state it upfront? One crucial issue that social anthropology raises in connection with the body is what it means to have already projected that concept in such a way that it appears as an object of knowledge. Part of the problem is its familiarity in the vernacular. As both Vilaça and Sharon Kaufman et al. (this volume) have shown us, it is a familiarity that equates body with a condition of the person understood as humanity. Part of the problem is how we construct objects of study, and I wish to exaggerate my usual crabwise posture of indirection in order to make that apparent. I therefore draw on a real-life situation in which I found myself talking about bodies in a manner I had not anticipated. As it happens, one way or another, my original excursus touched on several related themes that are also addressed in other chapters in this volume. Thus, like Vilaça for Amazonia, although less explicitly, my comments are informed by Melanesia. The two African ethnographers on which I draw themselves have Melanesian anthropology in view, and enable me to give some Melanesian reasons about why one might not wish to know in advance of analysis what 'the body' is going to convey.

Fractals and Fragments

It may well be a surprise to non-anthropologists that anthropologists of places such as Amazonia and Melanesia not only often fail to come up with descriptions of unitary communities, but cannot find unitary bodies either. Summoning 'the body' as a potentially totalized entity is a significant Euro-American artefact or image, of which we have a clear exposition in Maja Petrović-Šteger's chapter on Serbia. It (the idea that the body is recognizable as a form of 'the whole') is of course what makes one think 'fragments' might be a problem, though as Lesley Sharp (2005) has so pointedly described, more so for the state than the private person. Let's take two brief excerpts from African ethnography.

In his studies in Kabre (Togo, West Africa), Charles Piot (2005) describes the interdigitation of men's and women's bodily capacities in the household. So for example:

> a man's blood, that which allows him to cultivate and to produce children (the Kabre terms for blood and sperm are the same) comes largely, Kabre say, from the consumption of sorghum beer, and beer is the prototypical female product. Conversely, a woman's strength and womb – which allow her to work and produce children – come from the consumption of

porridge, which is produced from the prototypical male product (field crops). (2005: 69)

In other words, the wife's activities are already 'contained' within the husband's, and vice versa. It is not just that the division of labour renders the tasks interdependent but that the nature of the tasks (expenditure of bodily effort or ingestion of food from the soil) means that their bodies are interdependent, that the strength of each depends upon the strength of the other. Indeed Kabre say that what each produces belongs to the other: the food the husband produces is 'owned' by the wife and the children a woman produces are 'owned' by her husband. He calls this a state of reciprocal incorporation, and the same form of incorporation is repeated over and again.

One important locus is found in the way opposite conditions of being incorporate (or mutually constitute) each other 'at the level of the community'. All the houses in a Kabre community are divided into one of two clans, called male and female. The clan is a largely ritual entity; each is responsible for age-grade ceremonies in particular seasons of the year. Now amongst the male spirits of the male clan there also lives a female spirit for whom female ceremonies are performed, and among the female spirits of the female clan is a male one, in the same way that some houses of the one may retain their identity as having their origins in another. Each clan and its powers thus 'contains' the other within it. No single identity is totalizing.[1] Indeed it is no surprise that the ethnographer should talk of persons having multiple identities, such that a man from the female clan will act as a female on many occasions, as male on others, while his in-married wife (who is likely to come from the male clan) can appear male. Although a hierarchical ordering holds the male clan to be superior, the female clan being the one open to immigrants and the absorption of outsiders into the region,[2] either gender can appear encompassing, and encompassment always works to reinstate internal differences.

I offer these details to show that the notion of interdependence between bodies is part of a much wider system through which social identities of all kinds participate in one another. Bodies comprise a field of relations, we could say. But I also offer them because of the model that the anthropologist deploys. In organizing his account, Piot draws on an analogy with fractal mathematics. He is interested in the way in which forms repeat themselves – keep their form – throughout the social system while also being embedded in one another; thus the household with its internal division into male and female activities is contained within the community with its male and female clans.[3] The result is a series of self-similar,[4] non-linear symbolic orders, a 'fractal iteration'.

Piot's is one of two chapters on African materials in a volume on social anthropology and chaos (or complexity) theory. The second chapter (Taylor 2005), dealing with East Africa (Rwanda), shows us how some rather similar ideas illuminate what we may understand as communication between persons that also entail commmunication between bodies. 'The Rwandan person is fractal ... perennially incomplete. He or she is ever involved in the process of being added to, built upon, and produced by the gifts of others' (2005: 144). So, for instance, persons concerned to maintain the 'paths' of communication that lie between them enter into exchange relations and work to ensure the movement of people (e.g. women in marriage) and cattle (e.g. signifying alliances). Communication is imagined in terms of the flows and the blockages of bodily substances. Christopher Taylor emphasizes the analogy with the flow of liquids,[5] that have a fertilizing or productive effect. Thus the principal leader (king) in precolonial times, who controlled the passage of fluids, kept his own body 'open' through the quantities of milk, honey and mead that he took in and excreted; the conduit between earth and sky, the cosmos, was made manifest in his bodily state, and the health of the nation depended on him. The king could also, by withholding gifts and suchlike, block communication, impoverish as well as enrich his subjects. To summon a well-worn axiom (Lambert and McDonald, Introduction, this volume), what was true for the well-being of the polity was also true of the individual body.[6]

Fluids, says Taylor (2005: 145), are emphasized in popular medicine. When fluids are lacking or blocked within the body, or else haemorrhage out uncontrollably, one suspects illness and (or) sorcery. Of one study of healers and patients, he comments that flow/blockage symbolism suffused the narratives on both sides. This was in addition to whatever went on when patients consulted, as they also did, biomedical and other practitioners who operated on a different intellectual basis. Now what was happening to bodies was also happening to relations between persons – not only by force of analogy, as the observer might have it, but by the substantial connections through which (adopting Piot's vocabulary) the one contained or (adopting Wagner's [1991]) was a dimension of the other.

Here are 'parts', then, but the parts are not fragments of whole unitary identities; if they are parts of anything then we might say that they are parts of relations. So the fact that a male collectivity contains within itself a part that is female points to a relation between two kinds of origins or sources.

Multiplicity

Now the notion of plural or multiple identities is these days prevalent in Euro-American thinking. I take an innovative and challenging example already encountered in this volume. Anne-Marie Mol's (2002) book comes out of philosophy and science studies with the title *The Body Multiple*. Objects have complex relations, says Mol (2002: 149), and she means objects as they become the focus of people's dealings with one another, and hence of their knowledge. Thus in the course of an operation involving arteries, a patient's blood vessels ordinarily 'contained' within the body, in becoming the focus of attention become much larger than the body as a whole.[7] Not only is the body multiple: Mol is arguing that Euro-American practices reveal ways in which bodies are implicated in one another. Even if one remains entirely within Euro-American medical institutions, one can critique the fantasy of whole bodies that makes us so anxious about fragments. For powerful images do not necessarily make good analytics. Mol wants to work with the (analytical) concept of a multiple body by contrast with (the folk image of) a potentially fragmented and plural one.

Not only that: when the sociologist John Law (2004: 62) wants to make clear Mol's proposition about the multiple body he turns to the vocabulary of the fractional or fractal. Bodies are more than one but less than many.[8] However, I do not think that this 'multiplicity' of the philosopher's or sociologist's science studies argument is performed or enacted, and hence known, by the same conceptual techniques as inform gender divisions in (say) Kabre. The latter rest on the operation of division or partition between the sexes, such that one may also encompass 'parts' (taken) from the other. Roy Wagner's (1991) formula offers an abstract version of this technique, although not on the face of it in terms of bodies but in terms of persons and relations. The fractal person, he says, is the person with relations (to others) integral to it: neither singular nor plural, its dimensionality cannot be expressed in whole numbers (1991: 162). 'A fractal person is never a unit standing in relation to an aggregate … People exist reproductively by being "carried" as part of another' (1991: 163). What appears within one entity as a part of another entity is literally a part of the relation between them. This is the basis of the connection that I think Piot and Taylor are making between persons (in relations) and bodies (as parts of one another).

Indeed Melanesian materials would require me to put this more strongly. Mol's 'multiple body' is an image of overlapping elements, and does not have the bite of Melanesian 'partibility'. Whether we foreground persons, relations or bodies, partibility is premised on partition or division in the power to extract and absorb attributes and aspects of others. What is extremely interesting is the way in which Mol's argument is instead bound up with an argument about perspectives and contexts.

Perspectives and Contexts

We might remind ourselves of the Euro-American scientific view that holds that the physical world is continuous, and coterminous with the universe, the same for everyone, the only shortfall being in knowledge of it, such that diverse experiences are taken as a matter of diverse perspectives on it. Mol (2002) provides a wonderful rendition of the ontology of this perspectivalism, in which she shows how the diverse manifestation of objects in the world, their plurality, is produced by constantly shifting contexts of knowledge. Such shifts work insofar as contexts themselves are distinguishable from one another as relatively stable assemblages (Law 2004; Ong and Collier 2005).[9] Here I should add that I adopt Law's (2004) term 'perspectivalism' to distinguish it from that of 'perspectivism' initially deployed in Amazonia (e.g. Viveiros de Castro 1998). The terms are incidental, the conceptual difference profound.

Mol invites us to see perspectivalism thus. Suppose we take the contexts in which a disease is specifically produced by the instruments and methods through which it shows its effect. The accompanying assumption is likely to be that each such context is simply part of a larger whole, so all the perspectives brought to bear on a problem, or on an issue or concept, are thought to relate to one thing. People thereby manage, she says, to create a singular world out of the multiple ones of their behaviour and experience. By 'world' she means a contextually produced one, an epistemic object. What makes the account so interesting is that she demonstrates just how this 'one world' ('one nature' [Viveiros de Castro 1998]) of Euro-American epistemology is held together. Concomitantly all persons – imagined as 'human beings' – are in physical ('biological') terms inhabitants of this one world, and to that extent similar to one another. Their bodies are taken as evidence of this.

This holding together of diversity is of course very much an Enlightenment ontology. Her innovation is to deny both diversity and unity any primordial naturalism and show how they are 'performed' in the way people put their knowledge together. The inhabitants of this Euro-American world live with a unitary vision that is made to work against what they take to be a natural plurality, for as we have seen they understand multiplicity through the idea that the one world is also a plural one, composed of many elements. As a philosopher and ethnographer commenting on this world (Mol 2002: 7), Mol herself sees them as dealing with entities created by overlapping and intersecting fields or contexts of knowledge practice that have the character she analyses as multiple. Without wishing to criticize this profoundly illuminating exploration, the anthropologist might nonetheless remark on its limits.

It is clear that the term 'multiple' is not trivial in Mol's usage; it implies entities conjoined and disjunct from one another in ways that cannot be reduced to the folk notion of pluralities of parts gatherable into wholes. Her distinction between the plural and the multiple leads her to make a further, and radical, distinction between perspective and context. This is crucial to her critique of the (pluralist) perspectivalism that implies the priority of a unitary world on which or directed to which everyone has their own view; she adds that perspectivalism also creates a world of meaning from which the physicality of the body can disappear. 'Context' by contrast carries the force of an analytical tool in her account. Focusing on contexts – overlapping, complex, multiple – draws attention to the way the environs ('world') of an object are summoned and performed along with the manifestation or coming into existence of the object. And in the performance or practice, people enact their physical apprehensions of their experience.

But in the situation that I return to in a moment, the one that made me think about bodies at all, I had as much trouble with the concept of context as with perspective. The problem was both with the pluralism of perspective in the perspectivalist sense that Mol critiques *and* with contexts being seen to give rise to an apprehension of the world as multiply conceived. The remainder of the chapter will clarify this observation.

Communications

The convenors of the workshop on which the present volume is based invited us, through the title, to reflect on what it means to qualify 'bodies' by the epithet 'social'. Here I explore one diagnostic of the social, namely the communicative nature of interactions. Now Kaufman mentions the problematic ethics of clinical trials in developing countries (see Doumbo 2005), and at our original 2005 workshop, Hayley MacGregor went on to ask about the influence of interventions informed by anatomical models on the production of local knowledge about the body in illness (MacGregor 2005). The social processes of medical research, and paradigmatically clinical trials, were the focus of the real-life situation in which I found myself talking about bodies. As mentioned at the beginning, this was at a conference on the ethnography of medical research in Africa. It was held in the vicinity of a research unit of the Kenya Medical Research Institute in Kilifi District, and under the auspices of its senior staff. 'Research' for them involved both following through the global agendas of metropolitan research enterprises and conducting local trials on local populations carried out by largely Kenyan field researchers (e.g. Mbaabu, English and Molyneux 2005). All the practitioners were aware of the scope and limitations of the kind of knowledge they were bringing to their communications with participants (patients) in the course of undertaking medical research (Molyneux, Peshu and Marsh 2004; 2005).

I dub this knowledge Euro-American at source. Much Euro-American knowledge-making exists in the deliberate attempt to juxtapose contexts, that is, force one to *see* the difference it makes to actively displace, or simply eclipse, one set of aims or assumptions by another (Strathern 1995). This sense closes up the concepts that Mol analytically separates. 'Context' informs 'perspective' – perspective, that is, in the (perspectivalist) sense of a focus or viewpoint given by one's knowledge of things. Innovation or enhancement or movement is demonstrated in the difference itself. So for example the Nuffield Council on Bioethics (NCOB 2002) made a special case of biomedical research when it is carried out in developing countries. For conducting trials in a context where informed consent is a generally accepted ethical safeguard administered by medical practitioners becomes something else when one focuses on the participating population and their general expectations of medical practice.[10] Over time, the Council itself became aware of and wished to draw attention to a changing context of its own enquiry, namely the push to develop international bioethical protocols.[11] Here a deliberate shift of focus or perspective is revealed in the ability to switch from one context to another; the Council creates a sense of innovation in showing that the fresh context alters an already existing terrain. That one can shift knowingly across contexts becomes part of the knowledge that exists about any one (context). These common techniques lie behind Mol's and Law's disquisitions on Euro-American perspectivalism.

Contexts, and the way they make us aware of perspectives (and vice versa), in turn offer highly significant conceptual tools for apprehending social complexity, the interweaving and folding in on one another of diverse vantage points. They are so common, however, that the resulting manoeuvres may seem obvious and routine. In any event, they enable researchers or ethicists to approach, almost without thinking about it, the different rationales behind different inputs – the distinctive ends or aims of biomedical research, for example, put beside aims relating to immediate better health care for a particular population. Indeed, taking different viewpoints into account is such a standard assumption in the ethics and politics of consultation that it seems to flow seamlessly into equally standard assumptions about involving people in research.[12] On the one hand, the researcher needs to explain what he or she is about. On the other hand, there must be some measure of understanding on the part of the participant, who becomes 'informed' about a procedure, thereby acquiring new knowledge on the assumption that that new comprehension will replace or at least work beside old ones.[13]

This is all a matter of epistemology – a Euro-American one – with an interesting rider. For many medical researchers, I hypothecate, the issue is not just the desire to transmit knowledge but the idea that knowledge is a precondition to action, that understanding will inform practice. So, for example, they might imagine that someone who is told why a blood sample is being taken is more likely to be willing to accede than someone faced with an incomprehensible request. The request is justifiable in the light of the knowledge purveyed. The reverse also holds, that knowledge is being sought in order to tell one how to act. Indeed field trials often test previous attempts. In-so-far as trials are intervention-focused (looking at treatment efficacy), other kinds of research (clinical/microbiological/ epidemiological) into the relevant medical condition would have occurred at a prior stage. Field trials that attempt to ascertain the effectiveness of treatment in something approaching a 'real life' situation are already putting knowledge into action and are designed in turn to acquire knowledge about the effects of action itself (some kind of treatment intervention) so as to inform further action (a change in health policy, new clinical practice guidelines).[14] A new perspective makes the new aim intelligible, shifts people's perceptions of things, and an infinite number of such shifts may seem possible as new information becomes available. In the same way, the need to produce ethical principles or attend to internationally applicable guidelines becomes a source of new knowledge, a fresh context, for the practitioners/researchers.

Thus a switch of perspectives seems to be the obvious impetus to action. Consider again the NCOB report on developing countries. Methods must be devised, it says (2002: 74), for making sure that information will reach all members of a community. Given the existence of prior knowledge, an obvious 'method' is to offer a fresh perspective (my phrasing), for out of the presentation of viewpoints some kind of compromise can be struck. In discussing how researchers try to convey information in a comprehensible manner to participants, the report goes:

> One [way] is to incorporate local belief systems into the process of providing information. For example, the researchers might say: 'Although I as a doctor I believe that the disease is caused by germs … I understand that you believe it is caused by a demon. I respect that fact that you have this belief and I should like you to try this medicine to remove the disease. Removing

the disease is more important to us than whether we think it is caused by germs or a demon.' … [I]n some circumstances it will be possible to strike a balance between such a stance and the harnessing of local beliefs in the interests of improving participants' understanding of research. (NCOB 2002: 74–75)

Yet compromise does not work as automatically as that. We have an example to hand.

Resisting Enlightenment

The example I cite links up with the work of both Petrović-Šteger and Peers (this volume). When the UK Working Group on Human Remains (DCMS 2003) was preparing its report, it heard commentaries both from scientific and anthropological experts, and from representations from various 'communities' who wanted or were asked to put across their view of the matter. The task was to draw up guidelines to assist UK legislation for the tenure and repatriation of human remains currently being held in museums. This material ranged from skeletons to samples of tissue and specimens of bone and hair, but in one way or another comprised body parts.[15] They were – to refer to Sharp's (2005) work again – treated as fragments, that is, as parts of something else.

'Scientists' stood for those who regarded them as fragments of knowledge. Scientific interest in such materials was justified by the medical information that the remains were capable of yielding as well as by the information they held about human evolution. Who knew to what system of knowledge they would contribute? It was feared that knowledge would thus be lost with repatriation, since that might lead to their destruction by peoples who (for example) would wish to inter such remains. Such scientific knowledge was potentially useful, and to medicine in particular, a sense much enlarged by the recontextualization of interest in the materials afforded by molecular biology and developments in genetics. One could not foresee what new contexts might in the future render such materials valuable all over again. Statements from indigenous or first nations people in Australia, New Zealand and North America were put side by side with the scientific view. Here it was claimed, in a contemporary idiom to which their interlocutors responded, that the remains were fragments of a culture whose dignity had to be restored to a sense of holism or of people's families or lineages who in finding their ancestors again would find a new integrity. However, the point I wish to make is how the report was constructed in terms of a debate built up out of diverse discussions, as though these comprised two 'sides' each with its viewpoint or perspective.[16]

The assumption behind the document was that either party might attempt to persuade the other to see the force of the argument being put forward. One or other viewpoint, or a compromise position, might come to inform the action to be taken. There was some sense that experts and representatives (in written or verbal statements) were making different kinds of knowledge claims, that is, their perspectives came from their different social contexts. The scientists were largely experts representing the interests of universal knowledge about the human condition while the first nations peoples[17] were representatives of minority interests, expert in their own particular traditions or cultures. Nonetheless the discussion as reflected in the final report ultimately framed these as comparable if necessarily distinct perspectives. Thus in suggesting that one way forward might be independently supervised resolution procedures, the report notes among the benefits of this proposal that it would 'foster a mood of understanding among parties' (DCMS 2003: 157).

Yet this is in itself to take the viewpoint of one side, the Working Group trying to reach some kind of agreement. Not all the first nations representatives shared the grounding assumption here. From what they said, it appeared that the difference between the two sides (or rather, they and the government that was thinking about legislation, of which the Working Party seemed merely its instrument) could not be reduced to a difference of viewpoint or focus, in other words of a 'perspective' given by one's knowledge of things. It could not be reduced to the idea that with discussion and information they would appreciate the 'context' from which the scientists were operating, and shift their own viewpoint accordingly, with scientists shifting in the same way. Or that informing the Working Group of the context from which they were operating would be the basis of a compromise agreement.[18] Some of this was expressed in terms of refusal to acknowledge that their ancestors were in any sense scientific specimens – not 'also were' (their ancestors), as an additional perspective on other contexts they were prepared to contemplate.

For the difference between the two parties was not one of perspective, and was not to be understood by reference to context. To take the case of some of the Australian Aboriginal representations, they were also saying things that their interlocutors grasped much less easily than the appeal to holism or integrity. They said that they (the Aborigines) were *related* to their ancestors, the scientists were not. What was not immediately evident was the extent to which this apparently obvious fact had ground-shifting connotations. Or so I infer: what they were saying, I think, was that they were different kinds of people.[19] As different kinds of people their relationship to the human remains made the two parties, claims of a quite different order. To the Aborigines knowledge had brought them to this point, but there was nothing more they needed to know about themselves in order to press their claims, only a matter of proving how they were related, through dance, song and other evidences of entitlement. Knowing the conduct and meaning of such performances would, in turn, only be effective when deployed by those with the right to use them. That effectiveness could not be transferred through acquiring 'knowledge' as such. Ultimately, it was not a matter of knowing but of being.

We have here in microcosm a modelling of some of the ways in which people's understandings or states of being bypass one another. The interest for our purposes here is in the way the body or parts of the body appeared a vehicle for the relations being claimed – for the Aboriginal representatives, with respect both to their relations with the Australian and UK governments and to what they knew of scientific research.

It is an Enlightenment assumption that human beings, in inhabiting a plural world, are all similar to one another. What is the point of the counter-intuitive assertion of mine that the Australian Aborigines spokesmen (might have) regarded themselves as different kinds of people from museum scientists?[20] I was taking my cue and my language from Viveiros de Castro's extensive arguments for Amazonia perspectivism, and as Vilaça has presented in the previous chapter. They address an indigenous preoccupation with the interactions between persons when they appear to include both humans and non-humans (other animals); that appearance is entirely contingent on one's perspective, and the perspective is created by the body one has. These Amazonians hold that people (men, jaguars, and so forth) are different insofar as they exist in different worlds, and the different worlds are signalled by the fact that they inhabit different bodies. That is, their bodies are materially capable of different operations, effects, capacities, powers.[21] Let me stress that 'world' – 'universe' in Vilaça's terms (this volume) – is not a metaphor for context or perspective, but implies a specific physical habitat and in the Amazonian version the body that goes with such a habitation. Movement between such worlds is possible, under the right bodily regimen (through diet, discipline), that is. We cannot know in advance what constitutes a body in this kind of regime, and what operations done on the body imply, that is, the circumstances under which particular kinds of 'bodies' will emerge as an entity in persons' dealings with one another.

However, there is another response to the question about the counter-intuitive assertion that the Australian Aborigines' spokesmen (might have) regarded themselves as different kinds of people from museum scientists.[22] One could ask whether it was not simply the case that they had different kinds of relations to the remains. There is one sense in which this is true, although not in terms of the supposition that people are all similar but differentiated by their relations to one another. I would suggest rather that we might think of different relations as putting people into different worlds, or at any rate giving them different kinds of bodies. I have in mind certain kinship and gender systems. They do not have the Amazonian drama of seeing persons across (what is for Euro-Americans) a human-non-human divide, but they do involve the way people can or cannot have an effect on the bodily constitution of others.

In orienting oneself towards this set of persons rather than that, one may well become a different kind of being, in terms for example of the kind of dependency that is created between bodies.[23] The way mother's kin make bodily claims on a person can be very different from the effects that relations with father's kin have. In Mt Hagen (Papua New Guinea Highlands), being a sister or a wife is not so much a matter of context (perspective in the perspectivalist sense) as of the way in which a woman's health and bodily strength is sustained by her relations with others. She will look to her husband for preparing the land that yields daily sustenance, in the food grown on it that she eats, and that will help form the children she bears – what nourishes her will nourish them too. She is a conduit for the nurture than flows from the father to his child. But from the perspective (as in Amazonian perspectivism) of her natal kin what nourishes her will find its terminus in her. She cannot pass it on; her body blocks it. Rather than a conduit, for her natal kin her body is like an adornment for or extension of or a valuable from her own clan that is loaned to another for a while, but in the end must be returned to its origin. The mother's clan thus demand compensation, a return in the form of wealth, for what they have bestowed. It might be exaggerating to say that a married woman occupies different worlds in relation to these sets of kin;[24] she certainly occupies a different kind of body.

It has to be a matter of ethnographic enquiry the extent to which notions similar to those interpreted in this way from Amazonia or Papua New Guinea are replicated elsewhere. The extrapolations can only be in the manner of a trial. But they shift the grounds of anthropological knowledge sufficiently to make one alert to what one would wish to encompass within the term 'body'.

Emergent Issues

The staff at the Research Unit of the Kenyan Institute of Medical Research based in Kilifi were concerned that participants in their investigations into population health did not understand the research process; people grasped other aspects of the situation but not the nature of research as such. The staff went so far as to engage their fieldworkers in a study of the parents of the children in whom they were interested, for example, in defining mild malaria for future interventions or working out appropriate dose levels for anti-convulsants. Here is a fieldworker commenting on the understanding that parents had concerning extra blood taken on admission to a hospital ward:

> Many mums understand that the research blood will have an immediate benefit for them – and that is the reason they agree to it. They don't understand that it's a long-term benefit … mums just know that their children are given treatment. In fact I heard them complaining that too much blood is taken, which shows they don't understand … (Molyneux, Peshu and Marsh 2004: 2550)

Molyneux et al. (2004: 2553) observe:

> An apparent failure to explain research clearly and to adequately communicate clinical results to individuals and research results to the broader community has contributed to comments that blood samples may be collected and sold, possibly as transfusions … [Fieldworkers'] efforts to explain community level benefits, a particularly difficult message to get across, and media reports about the international trade in blood products – may also play a role. It is a plausible explanation of why an apparently wealthy institution goes to such lengths and expense to collect samples.

Fieldworkers' efforts to engage participants may be unusual in the emphasis they have given to the quality of understanding (Sharon Kaufman, pers. comm.), although they are only making explicit what is implied in the research process itself. Yet people seemingly protest at too much explanation. The following continues Molyneux et al.'s account (2004: 2552–53, original italics). Fieldworkers reported

> parents simply asking them to proceed with parents' concern about the child's condition, basic trust in hospital staff and in the Unit (*'they just say yes because they know people at the hospital have no bad intentions'*), and familiarity with KEMRI (*'before you've even finished they say, yes, I know about that from last time, it's fine'*).

The fieldworkers wanted people to see their viewpoint, and were also frustrated that they did not retain information.

These were some of the issues that informed the conference mentioned at the outset, comprising the field out of which the body or bodies emerged as an object of knowledge for me. This is not because 'bodies' were to the forefront of these issues (indeed ethnography from altogether elsewhere was offered on that), but because they have led me to a particular point about Euro-American knowledge practices. This point is the habitual recourse of researchers generally to the concept of 'perspective', in the (perspectivalist) sense of a focus or viewpoint given by people's knowledge of things, in accounting for their different positions, and to shifting 'contexts' of knowledge as a device to offer new information in place of old.

For among all the problems that accompany the field researchers' efforts are those that begin, not end, with the process of providing information. If one is outside a framework of shifting perspectives and remaking contexts of knowledge, the possibility of compromise as a compromise between viewpoints may not even be conceivable.[25] Whatever kind of research one was doing, it would not be sufficient to rely on the power of knowledge, to rely on displacement by context, so that one perspective becomes replaced by another. This is not just because words lack power, but because the epistemological techniques that make another perspective persuasive are not in place.

Let me put my proposition in the strongest possible terms.[26] Suppose there were people who did not invest in epistemology in the way that permeates Euro-American science and much general cultural practice besides. Suppose such people treated knowledge otherwise, and not as a matter of expertise and interest that distinguishes people, because (to paraphrase Viveiros de Castro 2003) basically everyone has the same kind of knowledge. That is, everyone learns from other people, and it is the relative power of and position of people, their condition or state of being, that distinguishes them. Knowledge can thus become a sign of that distinction, as when initiates are put through learning situations. But you cannot effectively use knowledge inappropriate to your condition, status, relations with others, any more than a Hagen woman can by will alter the effects of eating off her father's land from those that come from eating off her mother's land. You can move between different states of being[27] – and the different kinds of knowledge attendant on these – or even anticipate the states of being of others in their reactions to you, but this is more than a matter of adding new knowledge to old.

Understanding the place of knowledge in such systems might be useful when it comes to trying to impart new knowledge in order to effect change. However, on the face of it the forms of bodily communication such as those described for Kabre and Rwanda do not immediately depend on influencing people through transforming the perspectives from which they know things or their understanding of contexts. These are ontological rather than epistemic operations, for they affect people's conditions of being. Knowledge is a vector or aid, means rather than end. But then they are not concerned with change, with bringing about social transformation in people's lives, either. So how can all this be relevant to the efforts of the medical researchers? First, the contrast may help us to understand *them* (the researchers) better; second, we have learnt as much about communication as about bodies.

As to the first point, one target audience of the medical researcher's efforts, the medical practitioner, is also interested in deploying knowledge as a means to the ends of effective treatment. And, recalling Mol, he or she will do so through magnifying the body as the site of intervention. But magnification (or telescoping) is itself imagined as a matter of perspective or context, where the axiomatic object of knowledge is the phenomenal world perceived as an environment, a context par excellence for health and well-being. Kabre or Rwandan handlings of the body, by contrast, condense or disperse its elements from and to a network of persons; it is the condition of the ensuing relations that result in a person's health and sickness, which in turn affects others. What is outside the body can come, in this sense, not from the 'environment' but from other bodies. Now knowledge of the state, good or bad, of a person's relations is crucial; yet the knowledge is no basis for action without the substantial bodily conditions that make the person into the kind of being he or she is. Bodies emerge as an analogue of relations. For if relations are sustained by flows, interchanges, exchanges of objects of all kinds, so are bodies. Any one body is enmeshed in a network of bodies, and communication takes the form of flows (or blockage) between bodies. So individual bodies appear as parts rather than wholes, that is, as the implication of persons and relations in one another.

One can imagine, second, all kinds of outcomes for the perception of health and medical treatment, for relations between researchers and the researched, and indeed for the very course that the scientific research will take. The Kabre and Rwandan manipulation of the flows and blockages that once created certain specific kinds of bodies tells the anthropologist something, after all, about local practices of communication that may well have enduring value. For all the cases where it would be pointless looking for explicit parallels, there will be as many where the evidence – from these two ethnographies at least – indicates it would be highly profitable to pause before assuming we know what communication might entail.

An unusual aspect of the Kilifi research programme was its base in a hospital that could offer immediate benefit to the population in the vicinity. One more excerpt will underline how a perception of well-being becomes bound up with the participants' relations with the researchers. In fact people drew the medical teams into their lives by translating what the team was doing into notions of benefit for themselves:

> Primary reasons for consenting appeared to relate to study benefits rather than altruistic motives or a desire to contribute to an improved global understanding about a major public health problem. ... Three quarters

(78%) of those interviewed … said they could not or would not have refused the blood sample, usually because of the study benefits, or more general positive perceptions of the study team and of KEMRI: *'I'm used to it, why would I refuse now when they've been so good to us?'* (Molyneux et al. 2004: 2551, original italics)

In other words, while the writers are disappointed that at the end of it all so little understanding of the research aims had been imparted, the participants are in effect paying the researchers the compliment of saying that it is they who have brought good health to them.

To return to the 'social bodies' rubric of this volume. Suppose we take communication as a 'social' act par excellence. I have sketched some techniques for communication that involve bodily practices. It is possibly of interest to the anthropologist's comparative enterprise that they were prompted with reference not to Euro-American conventions about the body, but to Euro-American conventions about the elicitation of knowledge.

Postscript: Making Knowledge Relevant

If I thought I had found one way of creeping up on the concept of 'the body' unawares, that is, without attracting the attention of all the arguments that bristle round the subject, then I had until this was written been all too unaware myself of creeping up on something else. I bring forward my acknowledgements to the editors of the present volume (see following) for making me aware of a long-standing debate in a field virtually foreign to me, medical anthropology.

It is clear that anthropologists have been arguing endlessly with medical researchers and public health professionals about what they wish their (anthropological) knowledge to convey, and it intensifies with time: a recent, quite excoriating critique of anthropological failures at communication is given by Hemmings (2005). Yoder (1997), followed by commentators Bibeau (1997) and Coreil (1997), gives a penetrating analysis of one of the paradigms that dominate medical research in developing countries; Lambert (1996) addresses the same theme. We see here, first, the assumption that beliefs inform behaviour; second, therefore, that changes in belief must precede changes in behaviour; third, that in the short or long run scientific knowledge will displace cultural belief, bringing us, fourth, back to the position that knowledge (belief) leads to action (behaviour). The anthropologist is put in the position of providing information on those original, 'cultural', background beliefs. However, these authors argue, the paradigm is already shaky in being made explicit. And here is the fascinating anthropological observation for me, one that came out of a systematic evaluation of US health communication programmes: 'our analysis of survey data rarely found a close correspondence between changes in knowledge and changes in behaviour' (Yoder 1997: 132). Perhaps what this chapter offers is some commentary on the background beliefs of medical researchers and professionals that so strongly inform the paradigm that has it otherwise.

Yoder does of course argue that anthropologists should be far from content with being taken for cultural interpreters. They occupy the distinctive position of being able to offer social analysis, that is, social analysis of people's actions and behaviour, of their decision making and the constraints and possibilities entailed, and of how events are produced, in which belief or knowledge will play a part but with which there is no simple correlation. I think that some of the other papers in this present volume may show the way.

Acknowledgements

This chapter re-presents, in different form, a companion paper prepared for the conference *Locating the Field: the Ethnography of Medical Research in Africa* (convenors Wenzel Geissler and Sassy Molyneux, Kilifi, Kenya, 2005) under the title 'Can One Rely on Knowledge?' I am grateful to Wenzel Geissler's and Sassy Molyneux's inspiration, and to Norbert Peshu and Kevin Marsh for the hospitality of the KEMRI/Wellcome Trust Collaborative Research Programme at Kilifi. My thanks to the editors of this volume for their care and for their substantive questions and to Helen Lambert for some of the references.

In particular, an exemplary exercise published by Molyneux and her colleagues in a series of papers in *Social Science and Medicine* had articulated the need to develop a tool to explain the meaning of research based on concepts understandable in the general community. I am grateful to Sassy Molyneux for giving me copies of these papers and here draw heavily on one of them.

Notes

1. Vilaça (2005: 458) writing of Amazonia in fact observes, 'Alterity, not identity, is the default state'.
2. The highest-ranking ritual figure (or 'leader') is male, though this important man is also regarded as a 'child', and on occasion treated as a female,

encompassing both elements (both genders; both seniority and juniority) within himself.

3. The house itself is at once male (outside) and female (inside), each space incorporating cross-sex activity: a man's granary is located at the heart of the inner female household while a woman's grinding hut for preparing cereals is located in the male area outside the homestead.

4. '[R]ecursive self-similarity – the folding in of opposition upon opposition' (Piot 2005: 73). Linear models are found in arguments, for example, that do not see beyond gender dualisms and have to rely on the notion of mediation to explain encompassing orders.

5. Liquids include bodily fluids such as breast milk, semen and blood, and aliments such as milk, honey and beer; as well as rivers and rain. 'Liquids are employed symbolically as metonyms, parts of persons standing for and embodying the whole ... or as metaphors mediating between separate semantic domains' (Taylor 2005: 144).

6. The relation between the two could be described as holographic. 'Holography may be understood as a "fractional dimension" or dimensional "remainder" that replicates its figuration as part of the fabric of the field, through all changes of scale. Fractality, then, relates to, converts to and reproduces the whole, something as different from the sum as it is from the individual parts; a holographic form is thus "self-scaling"' (Wagner 1991: 166).

7. Nor does the patient 'contain' the body, which looms larger than the person's diverse social identities, which vanish or are overshadowed in turn by (say) the anxiety of relatives as the only part of the outside world to which the patient must respond. This is not quite the same as simply saying that there are numerous perspectives that can be brought to bear.

8. Note the inflection of singular and pural and compare Haraway's formulation, more than one and less than two: 'One is too few and two are too many' (1985: 65).

9. The contrast between [Euro-American] perspectivalism and [Amazonian and other] perspectivism was presented in Strathern (1999: 238ff.) in terms of infinite and finite perspective.

10. A concern with ethics and the rights people wish to exercise may turn into concern with the efficacy of the practice in question or for what it says of their relations to the government.

11. The shift can be seen between the first report (NCOB 2002), which addressed itself initially to external sponsors of medical research, and the follow-up discussion paper (NCOB 2005), which situated its review within a burgeoning international field of guidance and guidelines.

12. See, for example, the advice given to Mol (2002: 11) by a social scientist anxious that she realize how many perspectives she has to take into account in her study of atherosclerosis.

13. And beyond that, may be required to make decisions over their own treatment.

14. The above two sentences were suggested by the editors, to whom I most grateful.

15. I was a member of the Working Group; obviously I confine my remarks here to the published report.
16. The dual structure was deliberate, and is set out at the beginning of the report (DCMS 2003: 29 talks of 'an irreconcilable conflict between "scientists" and "indigenous people"', and of 'polarized views', which meant that consensus would be difficult, but it would be the Group's achievement to show that some was possible). The last chapter before the recommendations is called 'Resolving the Conflict'. This polarization meant, among other things, that the voices of indigenous scientists were taken as some kind of compromise position between opposed camps.
17. To adopt the North American convention for indigenous groups having minority status within an encompassing state.
18. E.g. to allow UK museums to retain portions of the materials for future DNA investigation, or in taking back any remains in question make them accessible to scientific enquiry. The scientists' compromise position is noted at DCMS (2003: 33).
19. Apropos the dual structure set out at the beginning of the report, the sentiment of irreconcilable conflict was nearer this position than that of polarized views. The report gives several verbatim comments on this viewpoint; e.g. 'It is our *direct* ancestors that are being experimented on' [original emphasis]; 'We went to Natural History Museum to see our ancestors and we were told that we cannot see them. For us it is like going to see somebody in hospital. To us the people in museums are not dead, they are living'; 'How can research possibly compare? We're tired of other people interpreting us to ourselves' (DCMS 2003: 55). (In some cases the recognition of the rightness of the claim seemed paramount over the desire for human remains to be handed over, although the claim could only be articulated in terms of such a demand; see DCMS 2003: 209, from the Foundation for Aboriginal and Islander Research Action.)
20. Law (2004) briefly raises the question, and after Verran (1998) he does so in talking about Aboriginal land claims, about persons existing in different worlds qua persons. (It is important to his rejection of the idea of ontological universalism.)
21. The worlds may not be visible to one another, so the bodies one sees belong to one's own world, while the form they take in other worlds will not be apparent at all. In this view, personhood extends across many animal including human species, who are differentiated by the bodies they inhabit rather than by their minds, thoughts, emotions or psychology. Vilaça (2005: 448) expatiates on the 'different bodies' attributed to Amerindians and whites.
22. Prompted by the editors, to whom thanks.
23. I have argued elsewhere that within a Melanesian kinship universe, say, acting out one set of relations means not only eclipsing another, but performing those relations from a specific perspective. (Thus when a son switches to being a sister's son, his father becomes his MB's affine [in-law].) In that reconfiguration, it might be too extreme to say that a new 'world' of kin relations appears, but that is the sense of it.

24. Although it has been observed, for example, that Melanesian persons may make themselves into new kinds of persons in order to grasp the opportunities of a new world (Hirsch 2001).
25. Though I could imagine a Kilifi patient seizing on the fact that a 'doctor' would *like* them to take the medicine, i.e. as the basis of a relationship.
26. I am grateful to Almut Schneider who drew me to this conclusion.
27. Including 'exchanging perspectives' (Strathern 1988), closer to the sense of (Amazonian) perspectivism than (Euro-American) perspectivalism.

References

Bibeau, Gilles. 1997. 'At Work in the Fields of Public Health: the Abuse of Rationality', special issue on *Knowledge and Practice in International Health, Medical Anthropology Quarterly* (n.s.) 11: 246–52.

Coreil, Jeannine. 1997. 'More Thoughts on Negotiating Relevance', special issue on *Knowledge and Practice in International Health, Medical Anthropology Quarterly* (n.s.), 11: 252–55.

DCMS. 2003. *The Report of the Working group on Human Remains*. London: Department of Culture, Media and Sport.

Doumbo, Ogobara. 2005. 'It Takes a Village: Medical Research and Ethics in Mali', *Science*, 307(5710): 679–81.

Geissler, Wenzel. 2004. 'Locating the Field: the Ethnography of Medical Research in Africa', Conference Proposal to the Wellcome Trust, London.

Haraway, Donna. 1985. 'A Manifesto for Cyborgs: Science, Technology, and Socialist Feminism in the 1980s', *Socialist Review* 80: 65–107.

Hemmings, C. 2005. 'Rethinking Medical Anthropology: How Anthropology is Failing Medicine', *Anthropology and Medicine* 12: 91–103.

Hirsch, Eric 2001. 'Making up People in Papua', *Journal of the Royal Anthropological Institute* (n.s.) 7: 241–56.

Junnarkar, Bipin. 2002. 'Sharing and Building Context', in D. Morey, M. Maybury and B. Thuraisingham (eds), *Knowledge Management: Classic and Contemporary Works*. Cambridge, MA.: MIT Press.

Lambert, Helen. 1996. 'Popular Therapeutics and Medical Preferences in Rural North India', *Lancet* 348: 1706–9.

Law, John. 2004. *After Method: Mess in Social Science Research*. London: Routledge.

MacGregor, H. 2005. 'Bodies in the Era of AIDS: the Case of South Africa', *Social Bodies* workshop, Girton College, Cambridge, 2005.

Lairumbi, G.M., C.S. Molyneux, R.W. Snow, K. Marsh, N. Peshu and M. English. 2005. 'Ethical Principles in Practice: Insights from Interviews with Researchers and Policy Makers in Kenya'. Presentation to Kilifi conference [see acknowledgements].

Mol, Annemarie. 2002. *The Body Multiple: Ontology in Medical Practice*. Durham, NC and London: Duke University Press.

Molyneux, C.S., N. Peshu and K. Marsh. 2004. 'Understanding of Informed Consent in a Low Income Setting: Three Case Studies from the Kenyan Coast', *Social Science and Medicine* 59: 2547–59.

———— 2005. 'Trust and Informed Consent: Insights from Community Members on the Kenyan Coast', *Social Science and Medicine* 61: 1463–73.

NCOB. 2002. *The Ethics of Research Related to Healthcare in Developing Countries,* London: Nuffield Council on Bioethics.

NCOB. 2005. *The Ethics of Research Related to Healthcare in Developing Countries: A Follow-up Discussion Paper.* London: Nuffield Council on Bioethics.

Ong, Aihwa and Stephen Collier (eds). 2005. *Global Assemblages: Technology, Politics and Ethics as Anthropological Problems.* Oxford and New York: Blackwell.

Piot, Charles. 2005. 'Fractal Figurations: Homologies and Hierarchies in Kabre Culture', in Mark Mosko and Fred Damon (eds), *On the Order of 'Chaos': Social Anthropology and the Science of Chaos.* New York and Oxford: Berghahn Books.

Sharp, Lesley. 2005. 'Re-Socialising the Technocratic Body', *Social Bodies* workshop, Girton College, Cambridge, 2005.

Strathern, Marilyn. 1988. *The Gender of the Gift: Problems with Women and Problems with Society in Melanesia.* Berkeley: University of California Press.

———— 1992. *After Nature: English Kinship in the Late Twentieth Century.* Cambridge: Cambridge University Press.

———— (ed.). 1995. *Shifting Contexts: Transformations in Anthropological Knowledge* (ASA 1993 Decennial Conference series). London: Routledge.

———— 1999. *Property, Substance and Effect: Anthropological Essays on Persons and Things.* London: Athlone Press.

Taylor, Christopher. 2005. 'Fluids and Fractals in Rwanda: Order and Chaos', in Mark Mosko and Fred Damon (eds), *On the Order of 'Chaos': Social Anthropology and the Science of Chaos.* New York and Oxford: Berghahn Books.

Verran, Helen. 1998. 'Re-imagining Land Ownership in Australia', *Postcolonial studies* 1: 237–54.

———— 2001. *Science and an African Logic.* Chicago and London: University of Chicago Press.

Vilaça, Aparecida. 2005. 'Chronically Unstable Bodies: Reflections on Amazonian Corporalities', *Journal of the Royal Anthropological Institute* (n.s.) 11: 445–64.

Vivieros de Castro, Eduardo. 1998. 'Cosmological deixis and Amerindian perspectivism: a View from Amazonia', *Journal of the Royal Anthropological Institute* (n.s.) 4: 469–88.

———— 2003. 'And'. After-dinner speech given at 'Anthropology and Science', 5th Decennial Conference of the ASA, Manchester: Manchester Papers in Social Anthropology, no. 7.

Wagner, Roy. 1991. 'The Fractal Person', in M Godelier and M Strathern (eds), *Big Men and Great Men: Personifications of Power in Melanesia.* Cambridge: Cambridge University Press.

Yoder, P. Stanley. 1997. 'Negotiating Relevance: Belief, Knowledge, and Practice in International Health Projects', special issue on *Knowledge and Practice in Internatinal Health, Medical Anthropology Quarterly,* (n.s.) 11: 131–46.

NOTES ON CONTRIBUTORS

Sharon R. Kaufman is Professor of Medical Anthropology in the Department of Anthropology, History and Social Medicine, the Institute for Health and Aging, and the Department of Social and Behavioral Sciences, University of California, San Francisco. Her research interests include medical technologies and their links to ethics and consumption, forms of life, and the ways in which 'risk society' is lived today. She is the author of *And a Time to Die: How American Hospitals Shape the End of Life* (2005).

Helen Lambert (D.Phil. Social Anthropology, Oxford) is Senior Lecturer in Medical Anthropology in the Department of Social Medicine, Bristol University. She has done fieldwork in both India and the UK. Her research interests include medical pluralism, gender, relatedness and the body in South Asia, public health, HIV prevention, and notions of evidence in anthropology and medicine. She has numerous publications in the anthropology of India and in medical anthropology; her most recent project was a Special Issue of *Social Science and Medicine* (2006) offering anthropological analyses of evidence-based health care.

Maryon McDonald studied Social Anthropology at Oxford University; she was previously Reader at Brunel University and since 1997 has been Fellow in Social Anthropology at Robinson College, Cambridge. Her research interests include nationalism, the EU, questions of accountability, medical anthropology and science and society issues; her fieldwork has been in France, the UK and in EU institutions. Her publications include a book, several papers and edited volumes on questions of identity, addiction and health, the anthropology of the EU and, most recently, *Languages of Accountability* (forthcoming with Berghahn). She is currently engaged on a large Leverhulme-funded project examining changing perceptions of 'the body'.

Laura Peers is Curator for the Americas collections at the Pitt Rivers Museum and a Reader in the School of Anthropology and Museum Ethnography at the University of Oxford. Her research interests include First Nations cultural histories and the meanings of historic objects to First Nations communities today. She served on the DCMS Working Group on Human Remains. Recent publications include *Playing Ourselves: Native American and First Nations Interpreters at Historic Reconstructions* (2007) and (with Alison Brown and members of the Kainai Nation) *'Pictures Bring Us Messages/Sinaakssiiksi Aohtsimaahpihkookiyaawa': Photographs and Histories from the Kainai Nation* (2006).

Maja Petrović-Šteger is a Research Fellow at Peterhouse College, Cambridge, and member of the Cambridge Department of Social Anthropology. Her research explores various contexts where bodies – whether living, dead, or in the form of medically usable remains – become the sites of economic, legal, political, scientific and artistic attention. She is completing a monograph on *The Measure of the Dead Body: Accounting for Human Remains in Postconflict Serbia and Tasmania.*

John Robb is Senior Lecturer in the Department of Archaeology, University of Cambridge. His research interests include archaeological theory, particularly of agency, the body and long-term change, Mediterranean prehistory, and human skeletal studies. His publications include *The Early Mediterranean Village: Agency, Material Culture and Social Change in Neolithic Italy* (2007).

Ann J. Russ is an anthropologist and senior instructor in the Department of Psychiatry at the University of Rochester Medical Center, USA. Her research has focused on aging and end of life in various clinical settings, including hospice and palliative care environments and, more recently, renal dialysis units. Her first-authored publications include '"Is There Life on Dialysis?": Time and Aging in a Clinically Sustained Existence' (*Medical Anthropology*, 2005) and 'The Value of "Life at Any Cost": Talk about Stopping Kidney Dialysis' (*Social Science and Medicine*, 2007).

Janet K. Shim is Assistant Professor of Sociology in the Department of Social and Behavioral Sciences at the University of California, San Francisco. Her research in medical sociology has focused on issues at the intersections of health inequalities, biomedical science and technologies, and race, gender and aging. Her publications include, among others, 'Clinical Life: Expectation and the Double Edge of Medical Promise' (*Health*, 2007), 'Risk, Life Extension, and the Pursuit of Medical Possibility' (*Sociology of Health and Illness*, 2006), 'Constructing "race" across the

Science-Lay Divide: Racial Formation in the Epidemiology and Experience of Cardiovascular Disease' (*Social Studies of Science*, 2005), and 'Biomedicalization: Technoscientific Transformations of Health, Illness, and U.S. Biomedicine' (*American Sociological Review*, 2003).

Marilyn Strathern (Ph.D. Cambridge) is currently Professor of Social Anthropology at the University of Cambridge and Mistress of Girton College. She has conducted fieldwork in Papua New Guinea and her publications include *The Gender of the Gift* (1988), *After Nature* (1992), the co-authored *Technologies of Procreation* (Edwards et al., 1993), *Property, Substance and Effect* (1999), the edited volumes *Audit Cultures* (2000) and *Transactions and Creativity* (Hirsch and Strathern 2004), and most recently *Kinship, Law and the Unexpected* (2005).

Aparecida Vilaça is Associate Professor of the Graduate Programme in Social Anthropology at the Museu Nacional of the Federal University of Rio de Janeiro (UFRJ), Fellow of the Guggenheim Foundation and Research Fellow, Level II, National Research Council. For twenty years she has conducted field research among the Wari' of the state of Rondônia (Brazil) and has published two monographs about them, *Comendo como gente. Formas do canibalismo wari'* (1992) and *Quem somos nós. Os Wari' encontram os broncos* (2006), as well as various articles in scientific journals in Brazil and abroad. She has been working on the themes of cannibalism, war, the body, cultural change and conversion to Christianity.

INDEX

human remains as kin and as data,
tensions between 79, 81
ideas of 52
kinship connectedness 38–40
kinship relations 6, 7, 10, 28, 31, 34,
48, 72, 136, 139
kinship studies 12n1
Melanesian kinship 167
'national' kinship 52, 70n1, 72n15
reciprocal relations of 92
social relations and patterns of 83–4
see also ancestors
Kleinman, Arthur 22
knowledge 149, 152, 153
Aboriginal knowledge 158–9
clinical knowledge 29
contexts of 153–4, 155
deployment of 156–7, 162–3
elicitation of 164
enlightenment, resistance to 157–60
ethical knowledge 19–21
Euro-American practices 155, 161–2
implications of 153, 156, 158–9,
162–3, 164–5
local knowledge, production of
154–5
perspective and 158
power of 162
as precondition to action 156
scientific knowledge 153
transmission of 156
universal knowledge 158
Koenig, Barbara 19
Konrad, M. 7, 12n5
Konrad, M. and Simpson, R. 12n3
Krmpotich, Cara 95
Kulišić, Špiro 70n2, 71n8

Lambek, M. 10–11
Lambek, M. and Strathern, A. 131,
144n13, 144n14
Lambert, H. and Rose, H. 4
Lambert, Helen 1–12, 40, 70, 80, 94, 95,
100, 105, 124, 151, 164, 171
The Lancet 109
Laqueur, Thomas 23, 38

Larson, Frances 95
Latour, B. 4, 12n4, 129, 136–7, 141, 142,
143
Latour, B. and Woolgar, S. 12n4
Law, John 149, 152, 153, 155, 167n20
Lazar, Stefan, Prince of Serbia 48–9
Leighton, M. 106
Leverhulme Centre for Human
Evolutionary Studies, Cambridge 89
Leverhulme Trust 1, 124
Lévy-Bruhl, L. 144n16
life
extension through renal
transplantation 17, 19, 22, 25, 26,
27, 29, 37, 38
and stages of life 18, 37, 39
Lima, T. 140, 142, 143n4
Lindman, J.M. and Tarter, M. 81, 86,
88
Lo Porto, F. 123
local knowledge, production of 154–5
Lock, M. and Scheper-Hughes, N. 130,
131
Lock, Margaret 2, 3, 4, 12n2, 22, 47,
130, 143n9
longevity 17, 18, 19, 26–7, 35
Lowmniasky, Henryk 70n2
Loy, Dr Tom 111–12, 115, 125n5

McCallum, C. 144n18
McDonald, Maryon 1–12, 40, 70, 80,
94, 95, 100, 105, 124, 151, 171
MacGregor, Hayley 154
Mandal, A.K. et al. 24, 26, 41n3
Maori remains 79, 80, 83
Marriott, M. 9
Marriott, M. and Inden, R. 9
Marsh, Kevin 165
Martin, E. 12n1
mass graveyards 56–9
materialism 102, 103, 104, 107, 117,
119, 120, 122, 124
Matthews, Maureen 95
Mauss, Marcel 18, 24, 92, 131
Mbaabu, L.G., English, M. and
Molyneux, C.S. 155

new reproductive technologies 2, 21, 27
New Zealand 157
Nolan, Marie T. 29
normalization of renal transplantation 19, 22, 23, 24, 28

Obeysekere, Gananath 93, 97n19
objectification 2, 26, 84, 132, 136, 144n13
obligation and gift exchange 24–7
O'Hanlon, Michael 95, 95n1
O'Neill, J. 2
Ong, A. and Collier, S. 153
ontology (and ontologies) 64, 106, 124, 142–3, 144n13, 162–3
  Amerindian ontologies 145n21, 145n22
  Enlightenment 153–4
  ontological divide 102, 130
  ontological instability 81, 96n9
  ontological rehabilitation of the sensible 131–2
  ontological universalism, rejection of 167n20
  perspectivalist 133, 140, 145n22, 153
organ donation *see* donation
organ recipients 6
  see also renal transplantation
organ transplantation 3, 5, 7, 10
  see also renal transplantation
Ötzi (the 'Ice Man') 100–101, 103, 107–9, 109–12, 113–17, 118–21, 122–4
Overing, Joanna 143n3

Parker, Ian 41n5
Parry, B. 3
Parsons, Talcott 23
Peers, L. and Brown, A.K. 96n7
Peers, Laura 3, 5, 6, 7–8, 77–97, 102, 104, 106, 117, 122, 172
perception 63, 136, 137, 139, 141, 163
  culturally differentiated 133
  embodiment and 131–2
  perspective and 156
personification of scientists 157–8

persons
  body as person 2, 3, 5, 9, 10, 11
  human remains as 47, 49, 52, 60, 66, 72n15, 73n21, 78–9, 81–2, 83, 85
  missing persons 52–3, 55, 56, 62–3, 64, 73–4n22, 73n16, 73n18
  personification of scientists 157–8
  repersonalization of Ice Man 101, 103–7, 112–18
perspectivalism 142, 153–4, 155, 160, 162, 166n9, 168n27
  perspectivalist ontologies 133, 140, 145n22, 153
perspective 148, 157, 158, 162, 163
  Amazonian 149–50, 153, 159–60, 166n9, 168n27
  appearance, body and 159, 166n7
  context and 153–4, 155–6, 159, 160
  First Nations peoples, perspectives of 157–8
  The 'Ice Man' (Ötzi), perspectives of 107–18
  knowledge and 158
  Melanesian 149–50, 152–3, 167n23
  perception and 156
  shift in 162
  theorization of 153–4
perspectivism 9, 130, 140, 152, 153, 159–60, 166n9, 168n27
Peshu, Norbert 165
Petrović, Sreten 70n2
Petrović-Šteger, Maja 6, 7–8, 47–74, 84, 104, 106, 117, 149, 172
Petryna, Adriana 60
Pfaffenberger, B. 120
phenomena
  phenomenology 2, 9, 27, 41n7, 131, 133, 136, 137, 139, 141
  pre-objective and genesis of 144n13
  renal transplantation as 'natural' phenomenon 21, 26, 34, 39
physics 4
Pinney, Christopher 96n9
Piot, Charles 11, 149–50, 151, 152, 166n4
Pitt Rivers Museum, Oxford 77, 78,